Springer Theses

Recognizing Outstanding Ph.D. Research

For further volumes:
http://www.springer.com/series/8790

Aims and Scope

The series "Springer Theses" brings together a selection of the very best Ph.D. theses from around the world and across the physical sciences. Nominated and endorsed by two recognized specialists, each published volume has been selected for its scientific excellence and the high impact of its contents for the pertinent field of research. For greater accessibility to non-specialists, the published versions include an extended introduction, as well as a foreword by the student's supervisor explaining the special relevance of the work for the field. As a whole, the series will provide a valuable resource both for newcomers to the research fields described, and for other scientists seeking detailed background information on special questions. Finally, it provides an accredited documentation of the valuable contributions made by today's younger generation of scientists.

Theses are accepted into the series by invited nomination only and must fulfill all of the following criteria

- They must be written in good English.
- The topic should fall within the confines of Chemistry, Physics, Earth Sciences, Engineering and related interdisciplinary fields such as Materials, Nanoscience, Chemical Engineering, Complex Systems and Biophysics.
- The work reported in the thesis must represent a significant scientific advance.
- If the thesis includes previously published material, permission to reproduce this must be gained from the respective copyright holder.
- They must have been examined and passed during the 12 months prior to nomination.
- Each thesis should include a foreword by the supervisor outlining the significance of its content.
- The theses should have a clearly defined structure including an introduction accessible to scientists not expert in that particular field.

Christopher Lane

The Development of a 2D Ultrasonic Array Inspection for Single Crystal Turbine Blades

Springer

Author
Dr. Christopher Lane
University of Bristol
Bristol
UK

ISSN 2190-5053 ISSN 2190-5061 (electronic)
ISBN 978-3-319-34510-9 ISBN 978-3-319-02517-9 (eBook)
DOI 10.1007/978-3-319-02517-9
Springer Cham Heidelberg New York Dordrecht London

Softcover reprint of the hardcover 1st edition 2014

Printed on acid-free paper

Springer is part of Springer Science+Business Media (www.springer.com)

Supervisor's Foreword

The turbine is at the heart of a jet engine and is spun at around 10,000 rpm by the high-pressure gas exiting the combustion system at temperatures in excess of 1,700 °C. The turbine generates a huge amount of power to operate the compressor and other auxiliary systems.

Turbine blades are wonders of engineering. Despite being only 100 mm in height, each blade can generate as much power as a Formula One car engine. They are made from a nickel-based, superalloy chosen for its toughness, low ductility and resistance to thermal creep. The blades are made in a special casting process where the material solidifies as a single crystal to eliminate fatigue initiation sites at grain boundaries. The crystallographic orientation angle is controlled so that it aligns to the blade axis to maximise creep resistance. The melting point of the blade material is 1,400 °C, which is some 300 °C lower than the surrounding gas temperature in the turbine. The blades are prevented from melting by a thin film of cool air that is drawn off from the compressor stage of the engine and pumped out through a system of tiny holes on the turbine blades. The turbine blades are high-value, safety-critical components operating in an extreme environment; therefore, assessing their integrity is vital. Prior to assembly into an engine, turbine blades are exhaustively inspected using a variety of techniques including dye-penetrant, ultrasound, eddy-currents and X-ray diffraction. Once in service, there is the potential for further defects to initiate, primarily due to thermomechanical fatigue. A crack initiating from fatigue is most likely to occur in the root section where the centrifugal loading is greatest and the complex geometry of internal cooling passages causes high stress concentrations.

This thesis describes the development of a new technique to inspect the root of turbine blades *in situ*, without dismantling the engine. The work—and the story told in the thesis—stretches all the way from the fundamentals of wave propagation in anisotropic material and ultrasonic array imaging through to device production and finally site trials. The thesis includes a description of a new method to determine crystallographic orientation from 2D ultrasonic array data. Another new method is described that enables volumetric images in an anisotropic material to be generated from 2D ultrasonic array data, based on measured crystallographic orientation. After extensive modelling, a suitable 2D array and deployment fixtures were manufactured and tested on *in situ* turbine blades in real engines. The final site trial indicated an order of magnitude improvement in the detectability of a certain type of root cracking over the best existing technique. The thesis should be an inspiration for those starting out on doctoral degrees as it shows the complete development cycle from basic science to industrial usage.

7 May 2013

Paul Wilcox
PhD (Lond), DIC Professor of Dynamics
Univeristy of Bristol

Contents

Chapter 1
Introduction

Abstract Turbine blades are the components within jet engines that provide the mechanical power for the compressor and auxiliary engine systems. They operate in extreme engineering conditions. Rotating at 10,000 rpm, the blades experience a tensile stress of 300 MPa whilst being surrounded by gas temperatures above 1,700 °C. Due to this demanding in-service operation, there is the potential for defects to initiate within the turbine blades. This could possibly lead to component failure and loss of engine power. This thesis details the development of a non-destructive evaluation (NDE) system for the inspection of turbine blades for potential defects. A successful NDE system could be used to mitigate the risk of in-service component failure by detecting and characterising the defects with high accuracy. The developed system is designed to perform the inspections in situ to eliminate the significant time and cost associated with removing the turbine blades from the engine prior to inspection. In this introductory chapter, the basic operation of jet engine turbine blades, their design and manufacture is briefly described. The motivation behind requiring high-sensitivity systems that can inspect the blades in situ for defects is explained. Next, the field of NDE is introduced and its importance to modern engineering, in particularly to the aerospace industry, is discussed. The main NDE methods are introduced and it is shown why an ultrasonic array inspection is chosen to be developed.

1.1 Motivation

A major new trend in the aerospace jet-engine industry is the increasing prevalence with which engines are sold to customers with *"power by the hour"* contracts. For example, in 2009 Rolls-Royce plc, the world's second largest engine manufacturer, had 90 % of its civil Trent engines under such contracts [1]. In a typical *"power by the hour"* contract, a fleet of engines is sold by the manufacturer

C. Lane, *The Development of a 2D Ultrasonic Array Inspection for Single Crystal Turbine Blades*, Springer Theses, DOI: 10.1007/978-3-319-02517-9_1,

with an agreement that ensures the operator will always have access to the fleet without having to purchase spare capacity. Therefore, the engine manufacturer agrees to supply spare engines, replace components, and provide all the required maintenance work whilst the customer pays a fixed price per operational hour. This benefits the customers as they have a predictable financial outgoing related to engine operation.

To ensure the profitability of these contracts, engine manufacturers must aim to increase the reliability and life of their components whilst maintaining an acceptable standard of safety. One way to achieve this is to periodically inspect engine components for defects that may have developed in service. The ability of these inspection systems to be used in situ, whilst the components are still inside the engine and the engine is on the aircraft, is highly advantageous. To demonstrate, the cost to strip a typical two-spool commercial engine, remove a set of high-pressure turbine blades, inspect them, and replace defective blades as required is in the order of £250,000. The time to perform the overhaul is approximately one month, during which the engine manufacturer is contractually obliged to provide a replacement engine. Therefore, there is a clear financial incentive to the engine manufacturer to have systems available to rapidly inspect engine components in situ to ensure they are defect free.

Currently, if a defect is detected in an engine component, the component must be replaced. However this may not necessarily be the requirement in the future. For example, the *damage tolerant design* philosophy is increasingly being used in the design of new engines and engine components, see Sect. 4.3 of Reed [2]. The basis of damage tolerant design is the assumption that all components contain defects. Defects range from intrinsic material discontinuities, to defects such as machining marks introduced during manufacture, and up to macroscopic defects such as cracks and corrosion that initiate in service [3, 4]. The size of the defects within each component is assumed to be just smaller than the size that can reliably be found during a post-manufacture and pre-service inspection, as any larger defects will be detected and the component scrapped. Stress and fatigue models are then used to predict the growth of the defects. Once the life of the component has reached a stage where the defect has potentially grown to a critical size, the component must either be withdrawn from service or inspected again for defects. The threshold is determined using fracture mechanics to predict the critical crack size that would cause fast catastrophic failure of the component, with an additional safety factor. Upon inspection, if the component has no defect indications, the component may remain in service. The assumed defect size in the component is then just below that of the inspection capability. Conversely, if a defect is found by the inspection, that component must be withdrawn from service. However, it is foreseeable that as stress and fatigue models are improved and validated, it may be permissible for a component that contains defects large enough to be detected but smaller than a critical size, to remain in service without jeopardising the component integrity. This presupposes that the inspection system is able to measure the defect size with sufficient accuracy. Therefore, an in situ inspection system should not just be able to reliably detect defects but also be able to size and characterise the defects too.

1.2 Thesis Objectives and Background

The work presented in this thesis is concerned with the development of an in situ inspection system for the detection of defects in jet-engine turbine blades. Specifically the inspection is for cracking in the root section of high-pressure turbine blades manufactured from single crystal nickel-based superalloy CMSX-4. The aim is to make the sensitivity of the inspection as high as possible in terms of defect detection and characterisation whilst being sufficiently practical to deploy as an in situ inspection. The capability of the developed system will also be evaluated quantitatively, to demonstrate how the system may be used in determining the overall component reliability in a statistically rigorous manner.

1.2.1 Single Crystal Turbine Blades

Turbine blades are the components within a jet-engine that extract work from the hot, high-pressure gas exiting the combustion system, providing shaft power for the compressor stages and auxiliary engine systems [5]. Figure 1.1 shows a schematic diagram of a two-spool turbo-fan engine. The high-pressure (HP) turbine is located closest to the combustor and is connected directly to the HP compressor. The compressor pressurises the gas flowing through the engine before it enters the combustion chamber. In a typical modern engine there will be two HP turbine stages: two sets of turbine blades on two separate turbine disks connected to the same HP shaft, and up to twelve HP compressor stages. The low-pressure (LP) turbine sits behind the HP turbine and is connected, via a separate LP shaft, to the LP compressor. In the case of a turbofan engine, commonly found on civil aircraft, the LP shaft is also connected to the fan, sitting at the front of the engine. Three-spool engines, such as the Rolls-Royce Trent family, have an additional intermediate-pressure (IP) stage between the HP and LP stages.

The power and efficiency of a gas-turbine engine is directly related to the temperature of the gas that enters the turbine. For example, the turbine entry temperature (TET) of Rolls-Royce civil jet-engines has risen 650 °C in the last 50 years to approximately 1700 °C [6]. In the last two decades thrust has increased by 60 % whilst fuel consumption has decreased by 15–20 %. One significant development that has enabled the increase in TET was the introduction of turbine cooling systems in the 1960s [5]. Previously the maximum TET in engines with uncooled turbine blades was approximately 1000 °C. In modern cooled engines, air at 650 °C is drawn off from the compressor and is forced through the cooling passages in the internal structure of the turbine blades. This cool air extracts heat from the blades and then emerges through surface holes on the blade to form a film of relatively cool air around the outer surface of the blade aerofoil. This film protects the blade from the gas that can be up to 300 °C hotter than the melting temperature of the turbine blade material (approximately 1400 °C [7]). Computer

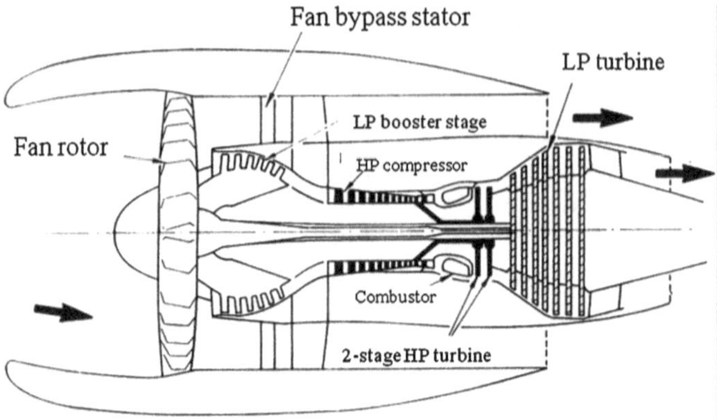

Fig. 1.1 Schematic diagram of a *two*-spool turbo-fan engine. Adapted from [5]

aided design (CAD) drawings of an example modern turbine blade are shown in Fig. 1.2. In these images the cooling holes are not shown but they would appear as perforations along the leading and trailing edge of the turbine blade aerofoil. The cooling system means modern turbine blades actually operate at a maximum temperature of 1050 °C.

However, at this operating temperature, creep becomes the significant determining factor for the integrity of the turbine blade, due to the centrifugal forces acting on the blades. Rotating at 10,000 rpm, a typical HP blade root experiences a tensile stress of 300 MPa, approximately 25 % of the ultimate tensile strength of the blade material. Resistance to creep at these loads and temperatures is achieved using nickel-based superalloys. Nickel-based superalloys exhibit excellent high temperature mechanical and environmental properties and have the ability to be precision cast [8]. Since the 1940s, nickel-based superalloys have been developed

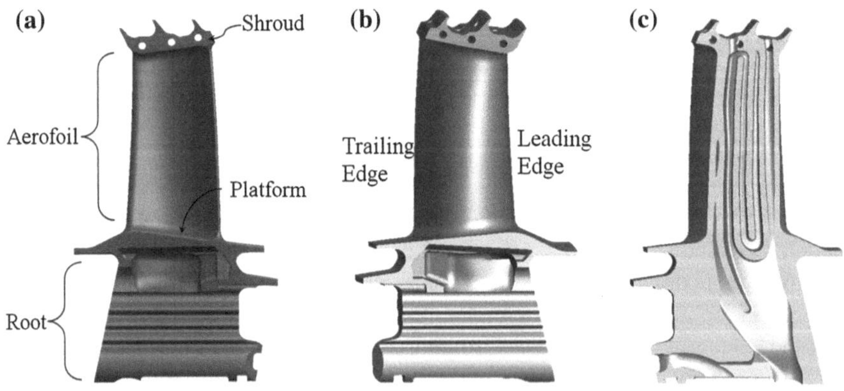

Fig. 1.2 Drawings of an HP turbine blade showing **a** the pressure side, **b** the suction side, and **c** a sectioned view showing the internal cooling passages. Images courtesy of Rolls-Royce plc

primarily for use in gas-turbine engines. Nickel was chosen because its face centre cubic (FCC) crystal structure means it is both tough and ductile and exhibits low rates of thermally activated creep. Furthermore, Nickel FCC structure is stable at elevated temperatures and therefore does not undergo phase transformation. Other metals with these crystal properties include the platinum group metals but these are dense and very expensive [2]. A major advancement in nickel alloys was the development of directionally solidified (DS) casting techniques in the 1960s. DS blades are cast with longitudinal grains aligned in the direction of the blade aerofoil. Therefore, grain boundaries are aligned with the stress axis, which removes the primary failure initiation sites. In addition, the low-modulus direction of the grains is aligned with the stress direction enhancing the thermal fatigue life, which is three to five times greater than conventional cast or equiaxed polycrystalline components [9].

A further development was the commercially viable manufacture of single crystal blades in the 1980s [10, 11]. Single crystals eliminate all grain boundaries which improves material properties further. This also allows the alloy composition to be optimised as grain boundary strengthening elements are no longer required. Table 1.1 shows the alloy composition of a single crystal nickel-based superalloy CMSX-4 developed by the Cannon-Muskegon Corporation [12]. In this alloy the aluminium and chromium enhance hot corrosion and oxidation resistance, and the additional alloying metals are used primarily as solid solution strengthening elements. This alloy is used widely for single crystal turbine blade manufacture and is the single crystal material that will be considered throughout this thesis.

The generic FCC crystal structure is shown schematically in Fig. 1.3. Eight atoms sit at the corners of a cube and six atoms are positioned at the centre of the cube's faces. Throughout this thesis the notation convention of describing crystal geometry [13] will be used, in which parentheses ( ) refer to individual planes and braces {} refer to a family of planes that are equivalent through symmetry. Likewise, brackets [ ] denote individual crystallographic directions and angle brackets $\langle\ \rangle$ are used for a family of equivalent directions.

In nickel-based superalloys the $< 001 >$ direction is the direction of lowest elastic modulus and highest creep resistance. This is also the natural solidification direction for the crystal. Therefore, single crystal turbine blades can be manufactured in a simple way to align the $< 001 >$ direction along the blade axis to achieve the optimal thermo-mechanical properties.

Single crystal turbine blades are manufactured using investment casting [2, 11]. Initially a wax model of the blade is made by pouring melted wax into a metallic mould containing a ceramic core that eventually forms the internal cooling passages. Individual wax models are then grouped together so that multiple turbine blades may be cast together. The wax assembly is coated in layers of ceramic slurry and stuccoed with larger ceramic particles. This process is repeated until the investment shell is sufficiently thick. After the ceramic coating is dry, the wax is melted away and the ceramic mould is fired. The mould is then placed in a vacuum furnace and molten alloy at 1550 °C is poured in. The bottom of the mould is placed on a water-cooled copper plate on which the alloy crystallises. The mould is

Table 1.1 Chemical composition of CMSX-4 by weight

Element	Nickel	Cobalt	Chromium	Tantalum	Tungsten	Aluminium	Rhenium	Titanium	Molybdenum	Hafnium
Composition (%)	61.7	9	6.5	6.5	6	5.6	3	1	0.6	0.1

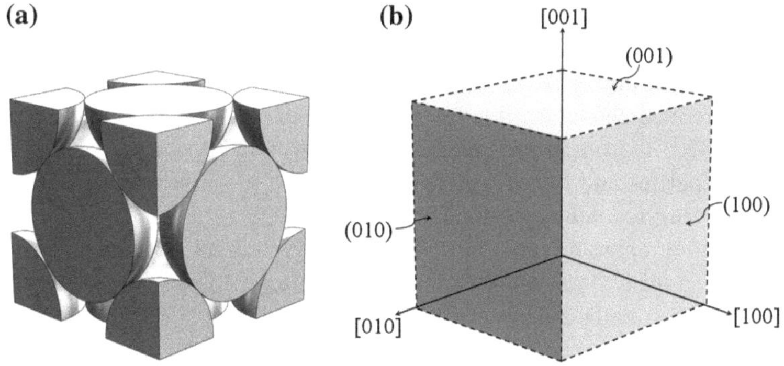

Fig. 1.3 **a** Schematic diagram showing the generic face centre cubic crystal structure. **b** A diagram showing the crystallographic geometry notation

withdrawn from the bottom of the furnace so that a temperature gradient forms that causes the crystals to grow up into the mould. Multiple crystals grow through the bottom of the individual blade moulds so a crystal selector must be used to ensure only a single crystal enters. The crystal selector works due to the fact that the $< 001 >$ direction is the preferred growth direction of the crystals. Grains with the $< 001 >$ direction aligned in the direction of the thermal gradient will grow the quickest. The crystal selector is a constricted passage, usually in a helical shape [14], in which the quickest solidifying crystal will branch and block off all other crystals. Once completely solidified, the investment shell is removed from the cast blades and the internal ceramic core is leached out with chemicals. The casting runner and risers are cut away before the single crystal turbine blades are machine-finished and the cooling holes are drilled using lasers.

It is estimated that over 1.5 million single crystal turbine blades that have been manufactured in this manner are currently in use in jet-engines [6].

1.2.2 NDE and the Inspection of Turbine Blades

Non-destructive evaluation (NDE) is a field of engineering that is concerned with investigating the integrity of a component in a manner that, as the name suggests, does not cause damage to the component. In general an NDE technique will inspect a component for macroscopic defects such as cracks or porosity. There are many NDE techniques available, each with their own capabilities and disadvantages. The majority of the most common techniques are used to inspect single crystal turbine blades, so a description of turbine blade inspection can also serve as an introduction to NDE.

After casting, the turbine blades are etched and visually inspected to check that no twin crystals exist. A twin crystal is a second crystal that has formed in addition to the primary single crystal. There is very little strength between the grain boundaries as there are no grain boundary strengthening elements in the single crystal alloy. To demonstrate, the presence of a secondary crystal in a CMSX-4 component with a difference in crystallographic orientation of 7° to the primary crystal can reduce the creep life of the component with an applied stress of 300 MPa at 850 °C from 10,000 h to 100 h [2]. A twin crystal will have different reflective properties and hence can be identified by a visual inspection. Turbine blades containing secondary crystals are scrapped.

The crystallographic orientation of the blades is then measured using an X-ray back diffraction method to ensure the crystal is correctly aligned with the geometry of the blade. This method is described in more detail later in Sect. 4.1.1.

An ultrasonic pulse-echo inspection is used to measure the wall thickness between the outer surface of the blade aerofoil and the internal cooling passages at a number of discrete points on the blade surface. Here an electrical pulse is applied across a piezoelectric element, within an ultrasonic transducer, causing the element to vibrate. The transducer is coupled to the turbine blade, and ultrasonic waves, typically with a frequency centred at 25 MHz, will propagate into the blade material. When the waves encounter the backwall of the cooling passage, the ultrasonic energy will be reflected. The piezoelectric element in the transducer converts the mechanical energy of the reflected ultrasonic wave back into an electrical signal. The time between the initial pulse and the reflected wave detected using the ultrasonic transducer is used to estimate the distance the wave pulse has travelled and hence the wall thickness of the turbine blade.

A fluorescent dye penetrant inspection (FPI) is performed to inspect the surface of the blades for casting defects. In a penetrant inspection a liquid dye is applied to the surface of the turbine blade. The dye is allowed time to seep into surface defects, through capillary action, and then excess dye is washed off. A developing chalk is then applied to the surface to draw the dye from any cracks. The specimen can then be visually inspected under UV light which causes the dye to fluoresce and highlight defect indications.

An X-ray inspection is then performed to check for internal defects such as inclusions, internal porosity and flashings. A flashing is where the metal alloy has solidified in cracks in the ceramic core during casting. Flashings can obstruct the flow of cooling air in the cooling passages during operation. The X-ray inspection is performed by penetrating the blade with ionising radiation. The radiation will be scattered by the material but a proportion will pass through and fall on X-ray film placed behind the blade. The film is processed to produce a 2D image that shows the intensity of radiation that has passed through the blade. Defects will cause a local change in the amount of radiation that penetrates through the blade, and hence defects are indentified as a variation in the local contrast of the X-ray image.

A common NDE technique not described above is eddy current inspection. An eddy current probe is constructed from a coil of wire that will produce an alternating magnetic field when an alternating current is applied to it. When the coil is placed in the vicinity of a conducting material, the primary magnetic field will induce an eddy current within the material. This eddy current will produce its own secondary magnetic field that will oppose the primary field and alter the impedance of the coil. A change in material properties or a defect within the specimen will perturb the eddy current, affecting the magnetic fields and changing the coil impedance. The amplitude and phase of the coil impedance can be measured to identify potential defects. Eddy current systems could be used for turbine blade inspection to detect surface breaking cracks and porosity. However, this inspection is achieved more accurately using FPI and therefore eddy current inspection is not used.

Once the turbine blades have been used in service there is the potential for further defects to initiate. This is primarily due to thermo-mechanical fatigue. During each engine cycle, in which an engine is started, used for take-off, cruise and landing and then shut down, the turbine blades experience a large range of operating temperatures causing fatigue. A crack initiating from fatigue is most likely to occur in the root section where the centrifugal loading is greatest and the complex geometry of internal cooling passages causes high stress concentrations. Hence, this project is concerned with inspecting for cracking in this region.

The root section of the turbine blades is inaccessible whilst the blades are in situ. Therefore, surface inspection techniques such as visual, FPI and eddy current inspections cannot be used. X-ray methods, which are able to detect internal defects, also cannot be used as there is limited access to fit X-ray film. Locations where the film could be positioned would require the X-rays to pass through high-density metal with a long path length which would cause the inspection sensitivity to be very poor. Therefore, the only practical inspection method for the in situ inspection of blade roots is ultrasound.

In addition to the ultrasonic wall thickness measurement described above, ultrasonic inspections may also be used to detect defects. A discontinuity in the elastic material, caused by a defect, will scatter any incident ultrasonic waves. These scattered waves can also be detected using an ultrasonic transducer to infer the presence of a defect. As detailed in the next section, ultrasonic *phased array* probes offer improved inspection capabilities over single element probes and will therefore be used for this project.

1.2.3 Ultrasonic Phased Array Inspections

Ultrasonic array probes consist of many individually wired ultrasonic elements within a single transducer package. The elements can be fired individually or in groups with different timings or *phase shifts*. This allows an array probe to steer and focus the ultrasonic beam in a way that would otherwise require a great number of different single element transducers. In addition, phased array probes are able to produce real-time images of the internal structure of a component, which enable components with complex geometries to be inspected with more accuracy.

Ultrasonic arrays were first used in the medical industry in the late 1960s for medical diagnostics. The first ultrasonic phased array consisted of 21 elements with an ultrasonic frequency of 1.3 MHz [15]. This medical array probe was shown to be able to successfully steer the ultrasonic beam by ±45°. Phased array probes for use in industrial NDE became increasingly common from the early 1980s, primarily for the inspection of components in nuclear power plants [16]. Currently, one-dimensional (1D) linear arrays, which consist of multiple elements spaced regularly along a single line are used extensively in industry, but they can only provide focusing and beam steering in a single plane [17]. However, 2D arrays, with elements arranged in a 2D plane, can manipulate an ultrasonic beam in every direction in the volume beneath the array, improving inspection coverage and sensitivity to angled reflectors [18–20].

As the turbine blades under consideration in this project have a complex internal geometry and the location and orientation of potential cracking in their root section can vary, 2D ultrasonic arrays will be developed to achieve the greatest volumetric coverage and the highest inspection sensitivity.

1.2.4 Anisotropy of Single Crystals

Single crystals are elastically anisotropic which means the elastic properties of single crystals vary with direction. For example, the elastic stiffness of a single crystal of pure nickel is 125 GPa in the $< 001 >$ crystallographic direction whereas in the $< 111 >$ direction it is 294 GPa [2]. Anisotropy also causes the propagation of ultrasonic waves in these materials to be dependent on the propagation direction [21]. Wave propagation in anisotropic materials is significantly more complex than in isotropic material and severely complicates the inspection of these materials with ultrasound. Therefore, the majority of this thesis is concerned with understanding the propagation of waves in anisotropic material and the subsequent compensation for the effects of anisotropy to ensure the highest possible inspection sensitivity.

1.3 Thesis Outline

Chapter 2, will begin with a discussion on the theory of wave propagation in anisotropic materials. Analytical models are developed that allow for ultrasonic array imaging algorithms to be corrected to compensate for wave propagation in anisotropic media. Two models are developed: the first computes the variation in velocity of ultrasonic waves propagating in different directions. The second computes the wave amplitude in different directions from a point-force acting on the surface of an anisotropic solid. This allows beam models to be produced to compute the ultrasonic field from an ultrasonic array element. The analytical models are validated using much slower numerical finite element models.

In Chap. 3, ultrasonic array imaging algorithms are presented and reviewed. The most appropriate algorithm for this project, the Total Focusing Method, is selected and is then corrected for imaging anisotropic components. Simulated data and experimental data are used to show the effectiveness of these corrections for 1D and 2D ultrasonic arrays. At the end of the third chapter, the effect of the misorientation of the single crystal on the calculated images is examined.

To reduce the effect of crystal misorientation, the crystallographic orientation of each blade should be measured. It will be shown that the most appropriate way to measure the crystallographic orientation is to use the same ultrasonic array that performs the defect detection. Therefore, Chap. 4 investigates potential methods to measure the orientation of a cubic single crystal using an ultrasonic array. Each suggested method is analysed using simulated and experimental data to determine its accuracy.

Chapter 5 details the development of a 2D ultrasonic array system for the in situ inspection of a single crystal turbine blade. The system is developed for a specific HP turbine blade with cracking in the root section. The corrected algorithms and orientation methods discussed in Chaps. 3, 4 are used to produce the most capable inspection system possible.

The 2D ultrasonic array system is analysed in a quantitative manner in Chap. 6 to determine the capability of the system in terms of defect detection and defect sizing.

Chapter 7, reviews the main findings of the thesis and suggests further relevant work that could be carried out.

As a stylistic point, each of the main chapters will begin with a literature review on the subject discussed in that chapter. The decision was taken to lay out the thesis in this way, rather than having a main literature review at the beginning of the whole thesis, due to the somewhat disparate topics in each of the chapters.

References

1. Annual Report 2009. 2010: Rolls-Royce Group plc
2. Reed RC (2006) The superalloys: fundamentals and applications. Cambridge University Press, Cambridge
3. Ball DL (2003) The role of nondestructive testing in aircraft damage tolerance. Mater Eval 61(7):814–818
4. Engine Structural Integrity Program MIL-HDBK-1783A. 1999: USA Department of Defense Handbook
5. Cumpsty N (2003) Jet Propulsion. Cambridge University Press, Cambridge
6. Ford T (1997) Single crystal blades. Aircr Eng Aeros Technol 69(6):564–566
7. Ricci E, Giuranno D, Novakovic R, Matsushita T, Seetharaman S, Brooks R, Chapman LA, Quested PN (2007) Density, surface tension, and viscosity of CMSX-4 superalloy. Int J Thermophys 28:1304–1321
8. Donachie M, Donachie S (2002) Superalloys: a technical guide, 2nd edn. ASM International, Ohio

9. Kong CN, High Temperature Mechanical Behaviour of Single Crystal Nickel-base Superalloys 2002, PhD Thesis, Department of Mechanical Engineering, University of Bristol Bristol, UK
10. Caron P, Khan T (1999) Evolution of Ni-based superalloys for single crystal gas turbine blade applications. Aerosp Sci Technol 3:513–523
11. Goulette MJ, Spilling PD, Arthey RP (1984) Cost Effective Single Crystals. In: 5th International Symposium on Superalloys, p 167–176
12. Frasier DJ, Whetstone JR, Harris K, Erickson GL, Schwer RE (1990) Process and alloy optimization for CMSX-4 superalloy single crystal airfoils. In: High temperature materials for power engineering, p 1281–1300
13. Guy AG (1960) Elements of physical metallurgy. Addison-Wesley, London
14. Carter P, Cox DC, Gandin CA, Reed RC (2000) Process modelling of grain selection during the solidification of single crystal superalloy castings. Mater Sci Eng, A 280:233–246
15. Somer JC (1968) Electronic sector scanning for ultrasonic diagnosis. Ultrasonics 6(3):153–159
16. Introduction to Phased Array Ultrasonic Technology Applications. Advanced Practical NDT Series. 2004, Waltham, Massachusetts: R/D Tech Corp
17. Drinkwater BW, Wilcox PD (2006) Ultrasonic arrays for non-destructive evaluation: a review. NDT & E Int 39(7):525–541
18. Velichko A, Wilcox PD (2010) Strategies for ultrasound imaging using two-dimensional arrays. Rev Prog Quant Nondestr Eval 29:887–894
19. Baba A, Kono N, Kitazawa S, Adachi Y, Odakura M, Kikuchi O (2010) Development of 3D phased array system “3D focus-UT”. Rev Prog Quant Nondestr Eval 29:941–947
20. Chatillon S, Mahaut S, Dubois P (2009) Simulation of advanced UT phased array techniques with matrix probes and dynamic settings for complex component inspections. Rev Prog Quant Nondestr Eval 28:864–871
21. Rose JL (1999) Ultrasonic waves in solid media. Cambridge University Press, Cambridge

Chapter 2
Wave Propagation in Anisotropic Media

Abstract Modern turbine blades are manufactured from single-crystal nickel-based superalloys for the excellent mechanical properties these materials exhibit at elevated temperatures. However, single-crystal materials are highly elastically anisotropic, meaning the elastic proprieties vary with direction. Wave propagation in anisotropic materials is significantly more complex than in isotropic materials. This severely complicates the inspection of these materials using ultrasound NDE methods. To enable the reliable inspection of single-crystal turbine blades with ultrasound, the propagation of ultrasonic waves in anisotropic material must be fully understood. In this chapter, the theory of bulk wave propagation in both isotropic and anisotropic solid materials is presented. Initially, a brief background of the research in wave propagation in solids is laid out to provide a historical context for this thesis. A short theory of bulk wave propagation is given and the important concepts of group and phase velocities are explained. Analytical models are then developed to describe wave propagation in anisotropic materials. Specifically, the developed analytical models simulate the variation in velocity with direction and the variation in beam amplitude from a point-force acting on the surface of an anisotropic material. These models are utilised in subsequent chapters to correct ultrasonic array imaging algorithms for the inspection of anisotropic materials. Finally, the analytical models are validated against numerical finite-element models.

2.1 Introduction

The history of research on wave propagation in anisotropic materials is linked with the development of the theory of elasticity in the early nineteenth century [1]. Early work by Green in 1839 showed that three wave modes could exist in a general anisotropic medium [2]. In 1877, Christoffel described the variation in the

C. Lane, *The Development of a 2D Ultrasonic Array Inspection for Single Crystal Turbine Blades*, Springer Theses, DOI: 10.1007/978-3-319-02517-9_2,

velocity of wavefronts in anisotropic media with the wavefront normal direction [3]. He developed mathematical formulae to compute this velocity, referred to as the phase velocity, using plane wave analysis.

The subsequent work on the theory of wave propagation in solids after this period was primarily concerned with and influenced by the fields of seismology and geophysics, with specific discoveries by scientists such as Rayleigh, Lamb and Love [4]. It has been suggested that as the elastic anisotropy of the Earth's crust is relatively small, little work was conducted on wave propagation in anisotropic solids during 1880–1910 [5]. Lord Kelvin began to investigate wave propagation in anisotropic media and reproduced Christoffel's findings in his Baltimore lectures in 1904 but concluded: 'the dynamics of elastic waves is a fine subject for investigation and I am sorry to pass from it now'[6].[1]

Very little progress was made until the 1950s when interest was revived due to the development of piezoelectric transducers and ultrasonic NDE methods. The ultrasonic inspection of anisotropic materials showed discrepancies between the theoretical knowledge and experimental findings. In the mid 1950s Musgrave [7] and Synge [8] used a wave energy based approach to compute the velocity of energy propagation in anisotropic solids which could then be shown to be equal to the group velocity in non-dispersive waves. In the 1960s, Merkulov developed a simplified derivation to compute the equation for the group velocity [9]. Merkulov also investigated the reflections of waves at the interfaces of an anisotropic solid and this was reviewed in more detail in 1968 by Fedorov [10].

In 1959, Buchwald was the first researcher to perform a detailed investigation of waves propagating from a point-force in an anisotropic elastic solid [11]. Using an asymptotic method developed by Lighthill [12], Buchwald derived an expression for the far-field amplitude of a wave excited by a point-force inside an infinite anisotropic solid. In 1971, Burridge used the Cagniard de-Hoop method to compute the amplitude of waves propagating from a point-force acting on the surface of an anisotropic half-space [13, 14]. Wu et al. developed a simplified approach to compute the far-field radiation for the same situation based on the reflection coefficients of waves at an interface of an anisotropic medium and the reciprocity theorem [15, 16].

The fundamental theory of wave propagation in anisotropic media thus is well understood and documented. For thorough discussions see [5, 17, 18]. Much of the current research in ultrasonic inspection of anisotropic components applies this theory to real applications and inspections. This will be discussed in the next chapter. What will be presented here is the theory of wave propagation that is required for this project, namely, the anisotropic velocity variation in bulk waves and the amplitude of bulk waves propagating from a point-force on an infinite anisotropic half-space. The analytical models are developed for both of these situations and are applied to CMSX-4 which is a single crystal nickel-based

---

[1] It may be of interest to note that Lord Kelvin did not approve of the prefix *an* for describing a non-isotropic solid. 'If we used this in English we should have to say: "An anisotropic solid is not an isotropic solid."' Instead Kelvin preferred the use of aeolotropic from the Greek *αιόλος* meaning variegated or different and *τρόποζ* meaning direction.

superalloy. The analytical models will be validated by comparing them to finite element models. The theory of surface waves or guided waves in anisotropic solids are not considered as they are not relevant to this thesis but references [17, 19] discuss these topics.

2.2 Bulk Waves in Anisotropic Solids

2.2.1 *Bulk Wave Propagation in Isotropic and Anisotropic Solids*

All metallic crystals are inherently anisotropic due to the directional differences in inter-atomic stiffness in the crystal structure. However, in conventionally cast engineering metals the microstructure is formed from many randomly orientated crystals, otherwise referred to as *grains*. In these polycrystalline materials, where the orientation of individual crystals is random, the mechanical properties on the macroscopic scale are independent of direction. Therefore, these materials are isotropic. In the case of ultrasonic wave propagation, when the ultrasonic wavelength is much greater than the grain size, isotropic assumptions are quite valid. Other metallic components, like single crystals or polycrystalline components, where the crystals in the microstructure material have a dominant orientation (for example directionally solidified castings, rolled materials or austenitic steel welds), exhibit highly anisotropic properties.

Bulk waves propagate in isotropic materials with equal velocity in every direction. Wave energy may propagate in two modes: longitudinal and shear. The faster mode is the longitudinal mode where the vibration of particles is *parallel* to the propagation of wave energy. The shear mode is the slower mode where the particle vibrations are *normal* to propagation direction. The direction of the particle vibration is referred to as the polarization.

The velocities, c_l and c_s, of the two bulk wave modes, where the sub-scripts l and s indicate longitudinal and shear modes respectively, can be computed using two elastic properties of the material, for example Young's Modulus E and Poisson's ratio υ, and the material density ρ using [20].

$$c_l = \left(\frac{E(1-v)}{\rho(1+v)(1-2v)}\right)^{\frac{1}{2}} \tag{2.1}$$

$$c_s = \left(\frac{E}{2\rho(1+v)}\right)^{\frac{1}{2}} \tag{2.2}$$

The wave propagation in anisotropic media is substantially different to the isotropic case. The most significant difference is that elastic waves propagate with a velocity that depends on direction. Furthermore, unlike the isotropic case, three wave modes can propagate: one longitudinal and two shear waves. However, these

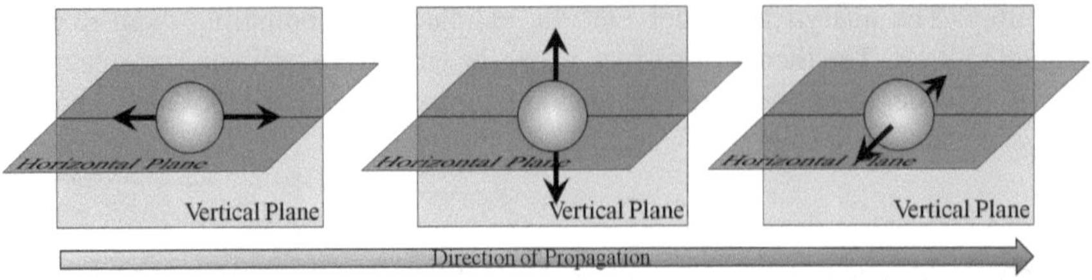

Fig. 2.1 Schematic diagram showing (from *left to right*) pure bulk wave propagation modes with longitudinal, shear-vertical and shear-horizontal polarization

modes are not necessarily pure modes as the particle vibration, or *polarization*, is neither parallel nor perpendicular to the propagation direction. In practice the anisotropic modes do show similarities to the isotropic modes and in these cases are referred to as quasi-longitudinal and quasi-shear. The quasi-shear modes are often distinguished further by whether they are primarily horizontally or vertically polarized. The polarization convention used for pure-modes is shown schematically in Fig. 2.1. The wavefronts of these quasi-modes generally do not lie normal to the energy propagating direction and hence the phase and group velocities do not coincide. These terms are explained in the next section. The phase and group velocities of waves in anisotropic solids can be computed from the material density and the elastic tensor that relates elastic stress and strain. In general there are up to 21 independent elastic constants in an anisotropic solid. However, most common engineering materials exhibit a lower degree of anisotropy and can therefore be described by fewer independent constants as shown in Table 2.1.

2.2.2 Phase and Group Velocity

The phase and group velocities are generally different in an anisotropic material and therefore a clear distinction between the two must be made to ensure the correct velocity profile is used when imaging with an ultrasonic array.

As a wave propagates in an elastic solid, particles within the material undergo a cyclic displacement. Particles at the same point within the cycle are said to have equal phase. Lines connecting particle of equal phase are referred as wavefronts. The velocity of the wavefront in the direction normal to the wavefront is known as the phase velocity. The spatial frequency of the wavefronts is referred to as the wavenumber. The wavenumber k of a wave propagating with phase velocity c and angular frequency ω is given by:

$$k = \frac{\omega}{c} \tag{2.3}$$

The group velocity is the velocity at which the envelope of a wave packet propagates. Figure 2.2 illustrates what is meant by the envelope of the wave packet.

Table 2.1 Anisotropic Materials

Material	Number of independent elastic constants	Engineering examples
Triclinic	21	–
Monoclinic	13	–
Orthotropic	9	–
Triagonal	7	–
Tetragonal	6	Yttria-stabilized zirconia for thermal barrier coatings
Transversely isotropic	5	Centrifugally cast stainless steel, welds
Hexagonal	5	Directional solidified castings
Cubic	3	Single crystal nickel castings
Isotropic	2	Polycrystalline metals

Auld [17] explains the concept of phase and group velocity mathematically by considering a modulated wave constructed of two waves of slightly different values of k and ω:

$$cos(kx - \omega t) + cos((k + \delta k)x - (\omega + \delta\omega)t) \tag{2.4}$$

Using the cosine rule, the equation describing the modulated wave is given by:

$$cos\left\{\left(k + \frac{\delta k}{2}\right)x - \left(\omega + \frac{\delta\omega}{2}\right)t\right\}cos\left\{\left(\frac{\delta k}{2}\right)x - \left(\frac{\delta\omega}{2}\right)t\right\} \tag{2.5}$$

The first cosine term in this equation is the high frequency carrier wave and the second term is the lower frequency modulation envelope. The velocity of the high frequency carrier wave can be seen to tend to the phase velocity as the small differences in ω and k tend to zero:

$$\frac{\omega + \frac{\delta\omega}{2}}{k + \frac{\delta k}{2}} \to \frac{\omega}{k} = c \tag{2.6}$$

The group velocity is the velocity of the lower frequency modulation envelope which, as the differences again tend to zero, is given by:

$$V = \frac{\delta\omega}{\delta k} \to \frac{\partial\omega}{\partial k} \tag{2.7}$$

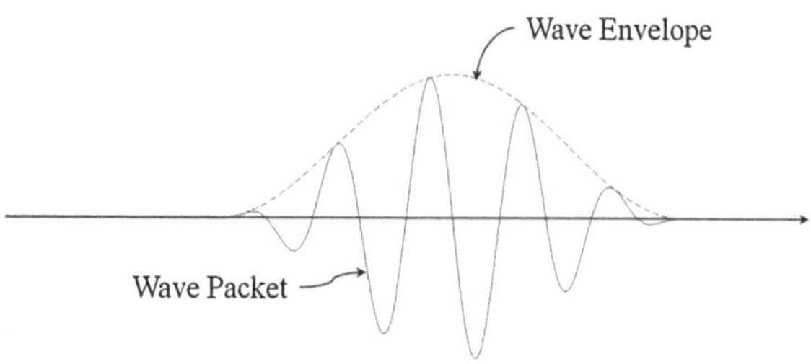

Fig. 2.2 Diagram showing the envelope of a wave packet

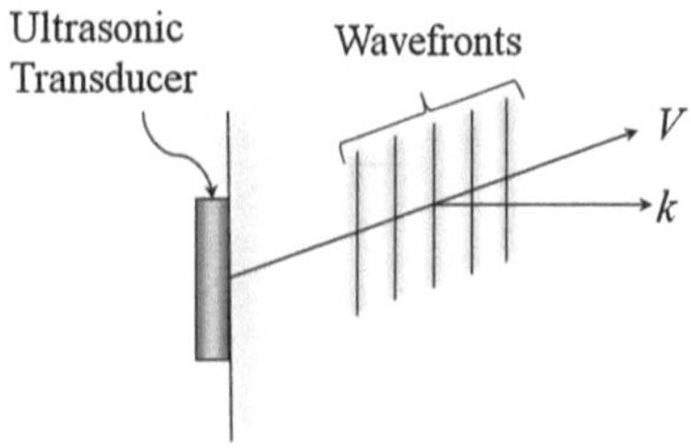

Fig. 2.3 Diagram showing plane waves propagating from an ultrasonic transducer in an anisotropic solid

In an anisotropic medium the physical difference between the phase and group velocity can be described by considering Fig. 2.3. The diagram shows a packet of plane waves propagating from an ultrasonic transducer. The wavefronts are parallel to the surface of the transducer and the velocities of the wavefronts propagate in the direction normal to the transducer surface given by the phase velocity. However, due to the anisotropy of material, the wave packet skews away from the normal direction and instead travels along the ray path. The velocity of the wave packet envelope is given by the group velocity.

Instead of the propagation of plane waves, the alternative case of a wave pulse propagating outwards from a point-source can also be considered. Figure 2.4 shows a plot of the wavefronts of the wave packet envelope propagating in an anisotropic solid from a point-source after a unit time. This plot is often referred to as the wavesurface and is equivalent to the plot of the group velocity as a function of ray path direction. In the propagation direction shown, the group velocity direction is clearly skewed from the wavenumber vector, which is always normal to the wavefront. Figure 2.4b shows how the group and phase velocities can be related through the skew angle ψ which can be expressed using the simple geometric equation:

$$c = V\cos(\psi) \tag{2.8}$$

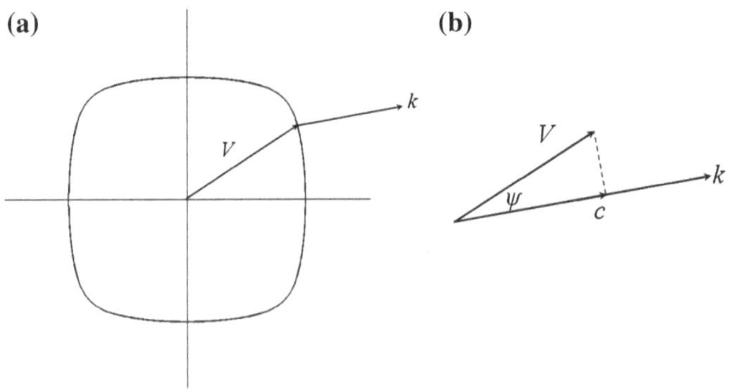

Fig. 2.4 **a** The quasi-longitudinal wavesurface of an anisotropic solid and **b** the geometric relationship between the group and phase velocities and the skew angle

2.2.3 Analytical Models

Analytical models to compute the phase and group velocities in an anisotropic material will now be derived. Tensor notation will be used for the derivation of these models for the compactness of this notation convention. In tensor notation, subscripts are used to show components of vectors in the three Cartesian coordinates, for example $x_i = \{x_1, x_2, x_3\}$. Einstein's summation convention is also used where variables with repeated subscripts are summed over all possible values of the subscripts (1, 2, 3) hence: $a_i b_i = \{a_1 b_1 + a_2 b_2 + a_3 b_3\}$. Differentiation with respect to direction is denoted using a comma, $u_{i,j} = \partial u_i/\partial x_j$, and therefore $u_{i,jk} = \partial^2 u_i/\partial x_j \partial x_k$. Differentiation with respect to time is indicated using a dot over the variable, $\dot{a} = \delta a/\delta t$.

2.2.3.1 Phase Velocity

The starting point to derive the equation for phase velocity in an anisotropic solid is Hooke's law. This law relates the stress σ acting on an elastic solid body to the strain ε. The stress at any point in a solid may be computed from the stresses acting on three mutually orthogonal faces. This can be described using the stress tensor. Tensors are often used to describe physical properties in a mathematical way so that the properties the tensor describes are invariant to a transformation in the coordinate system. The stress tensor, σ_{ij}, describes stress acting on the ith face in the jth direction of an elemental cube within the material. If the body is in equilibrium then the stress tensor must be symmetrical:

$$\sigma_{ij} = \sigma_{ji} \tag{2.9}$$

and hence the nine components of the tensor are made up of six independent elemental stresses. The elastic strain in a solid can also be described as a symmetrical tensor, ε_{ij}, and is related to the stress tensor through the elastic stiffness tensor C_{ijkl}:

$$\sigma_{ij} = C_{ijkl}\varepsilon_{kl} \tag{2.10}$$

The stiffness tensor is also symmetric:

$$C_{ijkl} = C_{jikl} = C_{ijlk} = C_{klij} \tag{2.11}$$

and hence the 81 (3^4) components can be reduced to 21 independent stiffness variables. The elastic constants of engineering materials are commonly expressed in Voigt notation as a symmetric 6×6 matrix:

Table 2.2 Indices contraction to swap between tensor and Voigt notation

Tensor indices	Voigt indices
$i = j = 1$	1
$i = j = 2$	2
$i = j = 3$	3
$i = 1, j = 2$ or $i = 2, j = 1$	6
$i = 1, j = 3$ or $i = 3, j = 1$	5
$i = 2, j = 3$ or $i = 3, j = 2$	4

$$C_{IJ} = \begin{bmatrix} C_{11} & C_{12} & C_{13} & C_{14} & C_{15} & C_{16} \\ - & C_{22} & C_{23} & C_{24} & C_{25} & C_{26} \\ - & - & C_{33} & C_{34} & C_{35} & C_{36} \\ - & - & - & C_{44} & C_{45} & C_{46} \\ - & - & - & - & C_{55} & C_{56} \\ - & - & - & - & - & C_{66} \end{bmatrix} \tag{2.12}$$

Therefore, the tensor notation must be contracted to the Voigt notation ($C_{ij,\ lk} \rightarrow C_{I,J}$) [18] and the appropriate indices swaps are shown in Table 2.2. Indice I in Voigt notation is calculated from tensor indices ij and likewise J from lk, so for example $C_{11,23}$ becomes $C_{1,4}$.

In the case of an isotropic material the elastic stiffness matrix in Voigt notation is:

$$C_{IJ} = \begin{bmatrix} C_{11} & C_{12} & C_{12} & 0 & 0 & 0 \\ - & C_{11} & C_{12} & 0 & 0 & 0 \\ - & - & C_{11} & 0 & 0 & 0 \\ - & - & - & C_{44} & 0 & 0 \\ - & - & - & - & C_{44} & 0 \\ - & - & - & - & - & C_{44} \end{bmatrix} \tag{2.13}$$

where:

$$C_{11} = \frac{E(1-\upsilon)}{(1+\upsilon)(1-2\upsilon)} \tag{2.14}$$

$$C_{12} = \frac{E\upsilon}{(1+\upsilon)(1-2\upsilon)} \tag{2.15}$$

$$C_{44} = \frac{1}{2}(C_{11} - C_{12}) = \frac{E}{2(1+\upsilon)} \tag{2.16}$$

The next step in computing the phase velocity is by using Newton's second law of motion to relate stress acting on a point in a solid to the motion of the particles in the solid. This law can be given in the form:

$$\rho \ddot{u}_i = \sigma_{ij,j} \tag{2.17}$$

where u_i is the displacement of a particle in the body in the ith direction. By differentiating Eq. (2.10) with respect to displacement, Newton's second law of motion and Hooke's law can be related through:

$$\rho \ddot{u}_i = C_{ijkl} \varepsilon_{kl,\,j} \tag{2.18}$$

The definition of elastic strain is:

$$\varepsilon_{kl} = \frac{1}{2}(u_{k,\,l} + u_{l,\,k}) \tag{2.19}$$

and hence Eq. (2.18) can be written as:

$$\rho \ddot{u}_i = \frac{1}{2} C_{ijkl}\left(u_{k,\,jl} + u_{l,\,jk}\right) \tag{2.20}$$

In tensor notation repeated indices may be swapped without changing the equation and therefore:

$$\rho \ddot{u}_i = C_{ijkl} u_{k,\,jl} \tag{2.21}$$

This is often referred to as Cauchy's law of motion. A solution to this equation can be assumed to take the form of a plane wave. This is useful because any wavefield can be described as a superposition of plane waves. The displacement of a particle at time t and distance x due to a one-dimensional plane wave can be described using:

$$u(x,t) = Acos(kx - \omega t) \tag{2.22}$$

where A is the wave amplitude. This is often expressed in exponential form using Euler's formula ($e^{ix} = cos(x) + i\ sin(x)$) as:

$$u(x,t) = Ae^{i(kx - \omega t)} \tag{2.23}$$

For a physical wave only the real part of this equation is used. A three-dimensional plane wave in tensor notation is therefore given by:

$$\ddot{u}_i = Ap_i e^{i(k_i x_i - \omega t)} \tag{2.24}$$

where p represents the polarization vector. To relate the plane wave solution to Eq. (2.21), the plane wave equation is first differentiated twice with respect to time:

$$\ddot{u}_i = -\omega^2 Ap_i e^{i(k_i x_i - \omega t)} \tag{2.25}$$

This is equivalent to:

$$\ddot{u}_i = -\omega^2 u_i \tag{2.26}$$

In a similar way, differentiating twice with respect to distance yields:

$$u_{k,jl} = -k_j k_l u_k \tag{2.27}$$

Therefore combining Eqs. (2.26) and (2.27) with Eq. (2.21) gives:

$$\rho\omega^2 u_i = C_{ijkl} k_j k_l u_k \tag{2.28}$$

The wave vector k_i is related to the wavenumber k through:

$$k_i = k n_i \tag{2.29}$$

where n_i represents the wavefront propagation direction. Therefore by substitution:

$$\rho\omega^2 u_i = C_{ijkl} k^2 n_j n_l u_k \tag{2.30}$$

Using the definition of phase velocity from Eq. (2.3) yields:

$$\rho c^2 u_i = C_{ijkl} n_j n_l u_k \tag{2.31}$$

This equation can be written in the form known as the Christoffel equation:

$$\left[\rho c^2 \delta_{ik} - C_{ijkl} n_j n_l\right]\left[u_k\right] = 0 \tag{2.32}$$

where δ_{ik} is the Kronecker delta:

$$\delta_{ik} = \begin{cases} 1, \text{ if } i = k \\ 0, \text{ if } i \neq k \end{cases} \tag{2.33}$$

The Christoffel equation can be seen to be an eigenvalue equation. Therefore, with the appropriate material properties, this equation can then be solved using eigenvalue analysis to compute the phase velocities of all possible wave modes with a particular wavefront normal (i.e. phase velocity direction) specified by n_j. The solution to the Christoffel equation yields three eigenvalues related to the phase velocity of the three wave modes that can exist in an anisotropic solid. Each eigenvalue has a corresponding eigenvector describing the displacement caused by each mode. The polarization vector of the individual wave modes is given by:

$$p_i = \frac{u_i}{\sqrt{u_j u_j}} \tag{2.34}$$

When the polarization vector is parallel to the wavefront normal direction the wave is a pure longitudinal mode. The wave is pure shear when the polarization vector is perpendicular to the wavefront normal direction. If the polarization vector fulfils neither of these conditions, the wave propagates in a quasi-mode.

2.2.3.2 Group Velocity

One possible way to obtain the group velocity profile is via a graphical method [13]. In this method the reciprocal of the phase velocity is plotted in the direction of the wavenumber vector as shown in Fig. 2.5. The resulting plot is referred to as

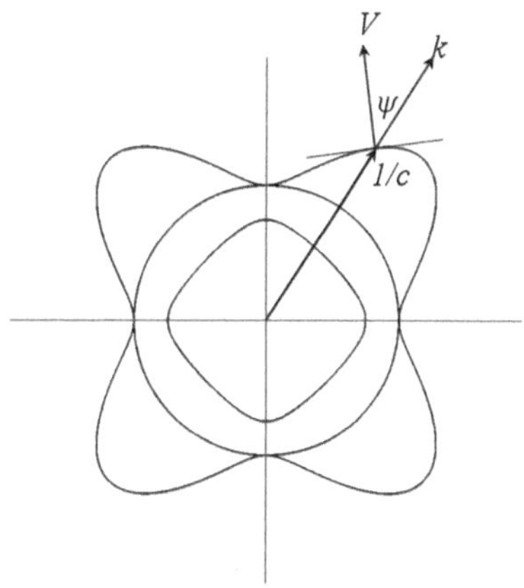

Fig. 2.5 Graphical method of computing the group velocity from the slowness surface

the slowness surface. The normal to the slowness surface is the direction of the group velocity. The angle between direction of the group velocity and the wavenumber vector yields the skew angle ψ. Hence, using Eq. (2.8) the group velocity can be computed.

However, practical implementation of this method is difficult due to numerical inaccuracies when measuring the normal to the slowness surface, especially as the slowness is a 3D surface. Therefore, an equation-based method is preferable. One approach to compute the group velocity is to consider the energy of a plane wave. These energy-based methods derive the energy velocity and then show how this is equal to the group velocity for lossless materials [10, 18]. The energy-based derivation is given in Appendix A. However, for the main body of this thesis, a simpler approach given by Merkulov [9] is used.

Starting from the definition of the group velocity in tensor form:

$$V_i = \frac{\partial \omega}{\partial k_i} \tag{2.35}$$

Now consider the Eq. (2.28) written using the polarization vector:

$$\rho \omega^2 p_i = C_{ijkl} k_k k_l p_j \tag{2.36}$$

Multiplying both sides by p_i yields:

$$\rho \omega^2 p_i p_i = C_{ijkl} k_k k_l p_j p_i \tag{2.37}$$

Because the polarization is a unit vector, ($p_i p_i = 1$), and therefore:

$$\rho \omega^2 = C_{ijkl} k_k k_l p_j p_i \tag{2.38}$$

Taking the derivative with respect to the wave vector, k_p, yields:

$$\rho \frac{\partial \omega^2}{\partial k_p} = C_{ijkl} p_j p_i \frac{\partial (k_k k_l)}{\partial k_p} \tag{2.39}$$

The right hand side of this equation can be solved using the product rule ($\partial(uv)/\partial x = u\partial v/\partial x + v\partial u/\partial x$) and the left hand side using $\partial u/\partial n = \partial u/\partial c \times \partial c/\partial n$ where $u = c^2$. Therefore:

$$2\rho\omega\frac{\partial\omega}{\partial k_p} = C_{ijkl}p_jp_i\left(k_k\frac{\partial k_l}{\partial k_p} + k_l\frac{\partial k_k}{\partial k_p}\right) \quad (2.40)$$

$$2\rho\omega\frac{\partial\omega}{\partial k_p} = C_{ijkl}p_jp_i\left(k_k\delta_{lp} + k_l\delta_{kp}\right) \quad (2.41)$$

Using the properties of the Kronecker delta:

$$2\rho\omega\frac{\partial\omega}{\partial k_p} = C_{ijkp}p_jp_ik_k + C_{ijpl}p_jp_ik_l \quad (2.42)$$

Repeated indices can be swapped, and therefore:

$$2\rho\omega\frac{\partial\omega}{\partial k_p} = C_{ijkp}p_jp_ik_k + C_{ijpk}p_jp_ik_k \quad (2.43)$$

$$\rho\omega\frac{\partial\omega}{\partial k_j} = C_{ijkl}p_lp_ik_k \quad (2.44)$$

Rearranging and using the definition of the phase velocity yields:

$$\frac{\partial\omega}{\partial k_j} = \frac{k}{\rho\omega}C_{ijkl}p_lp_in_k \quad (2.45)$$

$$\frac{\partial\omega}{\partial k_j} = \frac{C_{ijkl}n_kp_ip_l}{\rho c} \quad (2.46)$$

Finally, by referring to Eq. (2.35), it can be seen that:

$$V_j = \frac{C_{ijkl}n_kp_ip_l}{\rho c} \quad (2.47)$$

The group velocity is therefore the magnitude of the group velocity vector, V_j, which has its own propagation angle.

2.2.4 Results and Validation

The analytical models for waves propagating in a generic anisotropic material were applied to CMSX-4, the single crystal nickel-based superalloy considered throughout this thesis. The elastic properties of a cubic material such as CMSX-4 are given in Voigt notation as:

Table 2.3 Elastic constants and density of CMSX-4

Material property	Value at room temperature
C_{11}	235 GPa
C_{12}	142 GPa
C_{44}	131 GPa
ρ	8720 kg/m^3

$$C_{IJ} = \begin{bmatrix} C_{11} & C_{12} & C_{12} & 0 & 0 & 0 \\ C_{12} & C_{11} & C_{12} & 0 & 0 & 0 \\ C_{12} & C_{12} & C_{11} & 0 & 0 & 0 \\ 0 & 0 & 0 & C_{44} & 0 & 0 \\ 0 & 0 & 0 & 0 & C_{44} & 0 \\ 0 & 0 & 0 & 0 & 0 & C_{44} \end{bmatrix} \tag{2.48}$$

where $C_{44} \neq 0.5(C_{11} - C_{12})$. The elastic constants and density of CMSX-4 at room temperature are given in Table 2.3.

As Eq. (2.48) shows, the elastic properties of a cubic crystal can be described using just three independent variables, significantly less than the 21 variables needed to fully describe the most general anisotropic material. Therefore, in certain symmetry directions, simple relationships for the phase velocities can be achieved [21]. For example in the $< 001 >$ crystallographic directions in a cubic material the equations for the three wave modes are:

$$\rho c_{qL}^2 = C_{11} \tag{2.49}$$

$$\rho c_{qS_V}^2 = \rho c_{qS_H}^2 = C_{44} \tag{2.50}$$

where the subscript qL indicates the quasi-longitudinal modes, and qS_V and qS_H denote quasi-shear modes with vertical and horizontal polarization respectively. In the $< 011 >$ directions the phase velocities are:

$$\rho c_{qL}^2 = \frac{1}{2}(C_{11} + C_{12} + 2C_{44}) \tag{2.51}$$

$$\rho c_{qS_H}^2 = C_{44} \tag{2.52}$$

$$\rho c_{qS_V}^2 = \frac{1}{2}(C_{11} - C_{12}) \tag{2.53}$$

and finally in the $< 111 >$ directions:

$$\rho c_{qL}^2 = \frac{1}{3}(C_{11} + 2C_{12} + 4C_{44}) \tag{2.54}$$

$$\rho c_{qS_V}^2 = \rho c_{qS_H}^2 = \frac{1}{3}(C_{11} - C_{12} + C_{44}) \tag{2.55}$$

Fig. 2.6 Spherical Coordinates

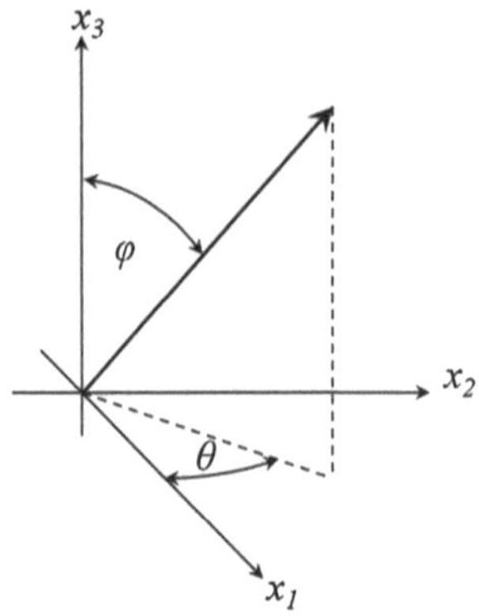

The above equations can serve as useful checks to ensure computer implementations of the analytical velocity models are correct.

To compute the velocity profiles of the three wave modes in all directions in CMSX-4, a computer program was developed in MATLAB 7.5.0 (The MathWorks Inc., Massachusetts, USA). The propagation directions were defined using spherical coordinates:

$$n_i = \begin{bmatrix} \cos\theta \sin\varphi \\ \sin\theta \sin\varphi \\ \cos\varphi \end{bmatrix} \tag{2.56}$$

where θ is the azimuth angle and φ is the zenith angle, as shown in Fig. 2.6. The complete range of propagation directions were computed with θ varying from -180 to 180° and φ varying from 0 to 180°. First the phase velocity was computed using Eq. (2.32) and the eigenvalue subroutines available in MATLAB. To check that the code had been implemented correctly, the computed phase velocities were compared to the close form solutions given in Eqs. (2.49) to (2.55). The phase velocities and polarization vectors computed from the eigenvalues and eigenvectors of Eq. (2.32) respectively were then used in Eq. (2.47) for the group velocity vector, from which the magnitude and direction of the group velocity could be computed.

Figure 2.7 shows the computed wavesurfaces (a plot of the group velocity vector) of the three wave modes for the CMSX-4 crystal in the {001} crystallographic plane. The quasi-longitudinal mode is labelled qL. The quasi-shear wave mode with horizontal polarization is labelled qS_H. In the {001} plane the mode is a pure-shear mode with isotropic characteristics. This is not the case for the other crystallographic planes. The second quasi-shear mode is polarized in the vertical plane and is labelled qS_V. In this plane the qS_V wave mode shows a highly anisotropic wavesurface. The points labelled A and B on the wavesurface are a conical point and cusp respectively. Between these points a cone in the wavesurface occurs where the qS_V wave has three different group velocities. Figure 2.8, a plot of the slowness surface from CMSX-4 in the {001} plane, shows graphically

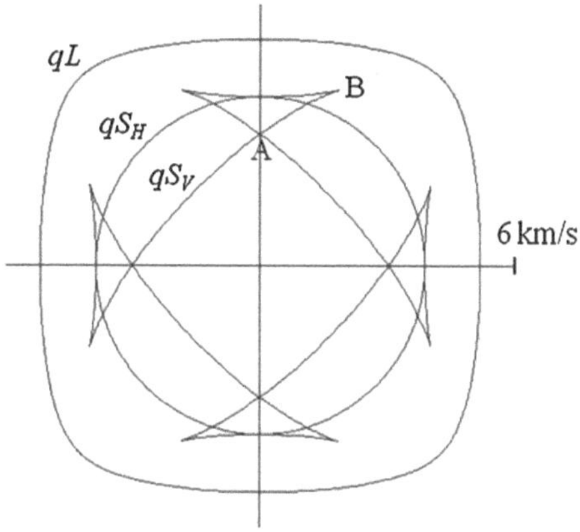

Fig. 2.7 The wavesurfaces of the three waves in single crystal CMSX-4. The plane shown is the {001} plane and the axes are aligned to the < 001 > crystallographic directions. The points labelled A and B on the vertically polarised quasi-shear wave shows a conical point and a cusp respectively

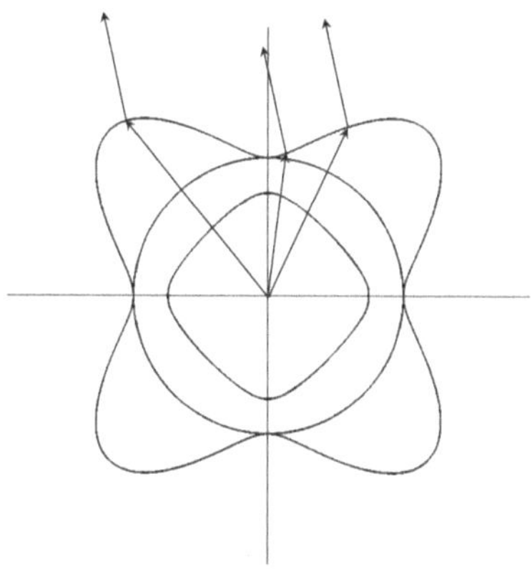

Fig. 2.8 A schematic diagram of the slowness surface of CMSX-4, showing how three separate wavefront directions can produce three different group velocities in a single group velocity direction

how three different phase vectors can produce three group velocity vectors in the same ray direction but with different magnitudes.

Figure 2.9 shows the quasi-longitudinal group velocity profile computed at all angles. This mode is shown in more detail as this is the mode that will be used later in the ultrasonic array imaging. The profile is shown as a spherical plot where the gray-scale represents the different velocities. The model predicts that the velocity varies from a minimum of 5.194 km/s in the < 001 > crystallographic directions to a maximum of 6.316 km/s in the < 111 > directions. The velocity profile can be seen to have reflective symmetry about the {001} planes.

2.3 Wave Amplitude from a Point-force on an Infinite Half-Space

Any arbitrary transducer can be modelled by the superposition of point-forces on the surface of the transducer. To be clear, a point-force refers to a force acting on a point in a single direction. These models can be used to compute the ultrasonic

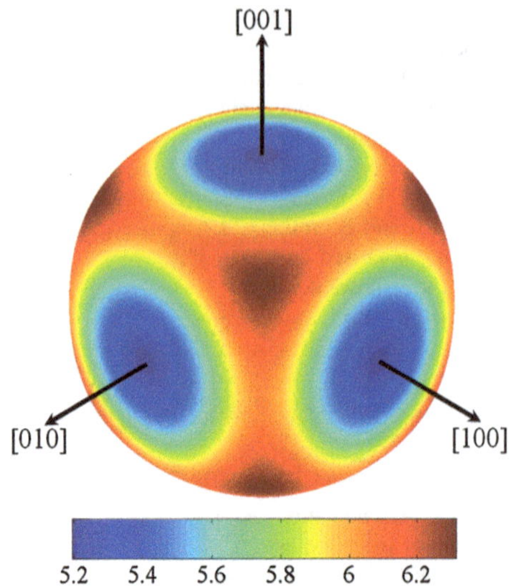

Fig. 2.9 A colourmap showing the *qL* group velocity profile in CMSX-4. Velocities are shown in units of km/s

field created by the transducer. This in turn enables the prediction of the amplitude response from a given reflector at different locations relative to the insonifying transducer in an anisotropic solid. These concepts will be discussed further in the next chapter. The model to compute the amplitude of a wave propagating from a point-force acting on the surface of an infinite anisotropic half-space in different directions will now be presented.

2.3.1 Analytical Model

The analytical model comprises two parts. The first uses the reciprocity theorem and reflection coefficients to compute the amplitude of waves propagating from a point-force acting on an anisotropic half-space due to Wu et al. [15]. The second part uses the anisotropic Green's function given by Buchwald [11], which describes the amplitude of a wave from a point-force in an infinite anisotropic as this is required as an input to the reciprocity calculations.

However, before the analytical models are presented it is often useful to rotate the elastic stiffness matrix to account for a rotation of the single crystal, or indeed any anisotropic material, relative to the surface on which the point-force is applied. This can be achieved using Bond matrix multiplication:

$$[C'_{IJ}] = [M][C_{IJ}][M^T] \tag{2.57}$$

where C'_{IJ} is the transformed elastic stiffness matrix, the superscript T indicates the transpose, and M is the Bond matrix given by:

$$[M] = [M_1][M_2][M_3] \tag{2.58}$$

The components of the Bond matrix can be given by three consecutive rotations θ_1, θ_2 and θ_3 about the three local Cartesian axes:

$$[M_1] = \begin{bmatrix} cos(\theta_1)^2 & sin(\theta_1)^2 & 0 & 0 & 0 & sin(2\theta_1) \\ sin(\theta_1)^2 & cos(\theta_1)^2 & 0 & 0 & 0 & -sin(2\theta_1) \\ 0 & 0 & 1 & 0 & 0 & 0 \\ 0 & 0 & 0 & cos(\theta_1) & -sin(\theta_1) & 0 \\ 0 & 0 & 0 & sin(\theta_1) & cos(\theta_1) & 0 \\ -1/2sin(2\theta_1) & 1/2sin(2\theta_1) & 0 & 0 & 0 & cos(2\theta_1) \end{bmatrix} \tag{2.59}$$

$$[M_2] = \begin{bmatrix} \cos(\theta_2)^2 & 0 & \sin(\theta_2)^2 & 0 & -\sin(2\theta_2) & 0 \\ 0 & 1 & 0 & 0 & 0 & 0 \\ \sin(\theta_2)^2 & 0 & \cos(\theta_2)^2 & 0 & \sin(2\theta_2) & 0 \\ 0 & 0 & 0 & \cos(\theta_2) & 0 & \sin(\theta_2) \\ 1/2\sin(2\theta_2) & 0 & -1/2\sin(2\theta_2) & 0 & \cos(2\theta_2) & 0 \\ 0 & 0 & 0 & -\sin(\theta_2) & 0 & \cos(\theta_2) \end{bmatrix} \tag{2.60}$$

$$[M_3] = \begin{bmatrix} 1 & 0 & 0 & 0 & 0 & 0 \\ 0 & cos(\theta_3)^2 & sin(\theta_3)^2 & sin(2\theta_3) & 0 & 0 \\ 0 & sin(\theta_3)^2 & cos(\theta_3)^2 & -sin(2\theta_3) & 0 & 0 \\ 0 & -1/2sin(2\theta_3) & 1/2sin(2\theta_3) & cos(2\theta_3) & 0 & 0 \\ 0 & 0 & 0 & 0 & cos(\theta_3) & sin(\theta_3) \\ 0 & 0 & 0 & 0 & -sin(\theta_3) & cos(\theta_3) \end{bmatrix} \tag{2.61}$$

Other rotation conventions can be successfully used, although appropriate changes to the Bond matrices would be required.

2.3.1.1 Normal Point-force on an Anisotropic Half-Space

The first step in the analytical model is to compute the wave vectors of the three possible reflected wave modes (one quasi-longitudinal and two quasi-shear) caused by an incident wave reflecting off a solid-void interface. From Snell's law, the projections of all the wave vectors on the interface surface must be equal and in the same plane as the incident wave (see Chap. 8 in Fedorov [10] or [22, 23]). In other words, the s_1 component of the slowness vector, s_i, is equivalent for all modes, as shown schematically in Fig. 2.10. In this example the incident and reflected waves are in the s_1–s_3 plane and therefore they have no s_2 component. Hence the remaining component of the slowness vector, s_3, can be found graphically by

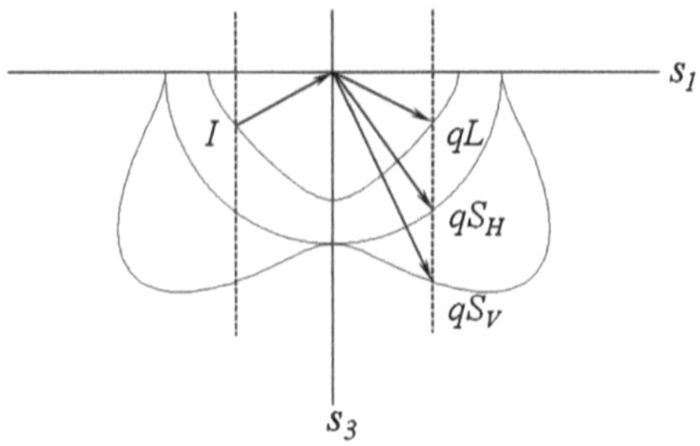

Fig. 2.10 Diagram showing the slowness surfaces of the three reflected wave modes from an incident wave

plotting the slowness surfaces of the three wave modes, and finding where the surfaces are intercepted by a line drawn at s_I.

To find the s_3 vector mathematically, the Christoffel equation is written in terms of the slowness vector:

$$\left|C_{ijkl}s_js_k - \rho\delta_{il}\right| = 0 \tag{2.62}$$

By defining G_{il} as:

$$\left|G_{il}\right| = \left|C_{ijkl}s_js_k - \rho\delta_{il}\right| \tag{2.63}$$

the terms can be expanded to give:

$$\begin{aligned} G_{11} = {} & C_{11}s_1^2 + C_{16}s_1s_2 + C_{15}s_1s_3 + C_{61}s_2s_1 + C_{66}s_2^2 + C_{65}s_2s_3 + C_{51}s_3s_1 \\ & + C_{56}s_3s_2 + C_{55}s_3^2 - \rho \end{aligned} \tag{2.64}$$

$$\begin{aligned} G_{12} & = G_{21} \\ & = C_{16}s_1^2 + C_{12}s_1s_2 + C_{14}s_1s_3 + C_{66}s_2s_1 + C_{62}s_2^2 + C_{64}s_2s_3 + C_{56}s_3s_1 \\ & \quad + C_{52}s_3s_2 + C_{54}s_3^2 \end{aligned} \tag{2.65}$$

$$\begin{aligned} G_{13} & = G_{31} \\ & = C_{15}s_1^2 + C_{14}s_1s_2 + C_{13}s_1s_3 + C_{65}s_2s_1 + C_{64}s_2^2 + C_{63}s_2s_3 + C_{55}s_3s_1 \\ & \quad + C_{54}s_3s_2 + C_{53}s_3^2 \end{aligned} \tag{2.66}$$

$$\begin{aligned} G_{22} = {} & C_{66}s_1^2 + C_{62}s_1s_2 + C_{64}s_1s_3 + C_{26}s_2s_1 + C_{22}s_2^2 + C_{24}s_2s_3 + C_{46}s_3s_1 \\ & + C_{42}s_3s_2 + C_{44}s_3^2 - \rho \end{aligned} \tag{2.67}$$

$$\begin{aligned} G_{23} & = G_{32} \\ & = C_{65}s_1^2 + C_{64}s_1s_2 + C_{63}s_1s_3 + C_{25}s_2s_1 + C_{24}s_2^2 + C_{23}s_2s_3 + C_{45}s_3s_1 \\ & \quad + C_{44}s_3s_2 + C_{43}s_3^2 \end{aligned} \tag{2.68}$$

$$\begin{aligned} G_{33} = {} & C_{55}s_1^2 + C_{54}s_1s_2 + C_{53}s_1s_3 + C_{45}s_2s_1 + C_{44}s_2^2 + C_{43}s_2s_3 + C_{35}s_3s_1 \\ & + C_{34}s_3s_2 + C_{33}s_3^2 - \rho \end{aligned} \tag{2.69}$$

Slowness vector component s_2 is set to zero to yield:

$$G_{11} = C_{11}s_1^2 + 2C_{15}s_1s_3 + C_{55}s_3^2 - \rho \quad (2.70)$$

$$G_{12} = C_{16}s_1^2 + (C_{14} + C_{56})s_1s_3 + C_{54}s_3^2 \quad (2.71)$$

$$G_{13} = C_{15}s_1^2 + (C_{13} + C_{55})s_1s_3 + C_{53}s_3^2 \quad (2.72)$$

$$G_{22} = C_{66}s_1^2 + 2C_{64}s_1s_3 + C_{44}s_3^2 - \rho \quad (2.73)$$

$$G_{23} = C_{65}s_1^2 + (C_{63} + C_{45})s_1s_3 + C_{43}s_3^2 \quad (2.74)$$

$$G_{33} = C_{55}s_1^2 + 2C_{53}s_1s_3 + C_{33}s_3^2 - \rho \quad (2.75)$$

Therefore, any given s_1 yields a sextic equation which can be solved for s_3. This sextic equation will give six slowness vectors for any s_1 input. However, the solutions that are valid are ones that cause the energy to be reflected back into the solid and inappropriate solutions are discarded.

Having computed the slowness vector of each reflected wave mode, the polarization, p_i, of the three waves must also be found. This is achieved by inserting the slowness vector values into the modified Christoffel Eq. (2.62) and computing the eigenvectors to give the wave polarizations.

Next, the amplitudes of the reflected waves can be found by considering the stresses at the interface. The stress caused by a propagating wave can be expressed as:

$$\sigma_{ij} = C_{ijkl}\varepsilon_{kl} = C_{ijkl}\frac{du_k}{dx_l} \quad (2.76)$$

Again, assuming the solution takes the form of a plane wave and differentiating the equation of the plane wave (2.24) with respect to direction yields:

$$\sigma_{ij} = C_{ijkl}Ap_lk_lu_k \quad (2.77)$$

The stresses of the incident wave and the three reflected waves that have a component normal to the interface must balance. These are σ_{33}, σ_{13} and σ_{23} and hence:

$$\sigma_{33}^{I} = \sigma_{33}^{qL} + \sigma_{33}^{qS_H} + \sigma_{33}^{qS_V} \quad (2.78)$$

$$\sigma_{13}^{I} = \sigma_{13}^{qL} + \sigma_{13}^{qS_H} + \sigma_{13}^{qS_V} \quad (2.79)$$

$$\sigma_{23}^{I} = \sigma_{23}^{qL} + \sigma_{23}^{qS_H} + \sigma_{23}^{qS_V} \quad (2.80)$$

Here the superscripts I, qL, qS_V, qS_H refer to the incident wave, the reflected quasi-longitudinal wave and the reflected quasi-shear waves with vertical and horizontal polarization respectively. There must also be continuity of the displacement vectors at the surface of the half-space. Therefore combining Eq. (2.77) with Eqs. (2.78) to (2.80) and removing the constant variables gives:

$$C_{33kl}A^I p_l^I k_k^I = C_{33kl}A^{qL} p_l^{qL} k_k^{qL} + C_{33kl}A^{qS_H} p_l^{qS_H} k_k^{qS_H} + C_{33kl}A^{qS_V} p_l^{qS_V} k_k^{qS_V} \tag{2.81}$$

$$C_{13kl}A^I p_l^I k_k^I = C_{13kl}A^{qL} p_l^{qL} k_k^{qL} + C_{13kl}A^{qS_H} p_l^{qS_H} k_k^{qS_H} + C_{13kl}A^{qS_V} p_l^{qS_V} k_k^{qS_V} \tag{2.82}$$

$$C_{23kl}A^I p_l^I k_k^I = C_{23kl}A^{qL} p_l^{qL} k_k^{qL} + C_{23kl}A^{qS_H} p_l^{qS_H} k_k^{qS_H} + C_{23kl}A^{qS_V} p_l^{qS_V} k_k^{qS_V} \tag{2.83}$$

Grouping terms of unknown amplitudes gives the following:

$$\begin{pmatrix} C_{33kl}A^I p_l^I k_k^I \\ C_{13kl}A^I p_l^I k_k^I \\ C_{23kl}A^I p_l^I k_k^I \end{pmatrix} = \begin{pmatrix} C_{33kl}p_l^{qL} k_k^{qL} & C_{33kl}p_l^{qS_H} k_k^{qS_H} & C_{33kl}p_l^{qS_V} k_k^{qS_V} \\ C_{13kl}p_l^{qL} k_k^{qL} & C_{13kl}p_l^{qS_H} k_k^{qS_H} & C_{13kl}p_l^{qS_V} k_k^{qS_V} \\ C_{23kl}p_l^{qL} k_k^{qL} & C_{23kl}p_l^{qS_H} k_k^{qS_H} & C_{23kl}p_l^{qS_V} k_k^{qS_V} \end{pmatrix} \begin{pmatrix} A^{qL} \\ A^{qS_H} \\ A^{qS_V} \end{pmatrix} \tag{2.84}$$

Therefore, for a given incident wave the amplitudes of the reflected waves can be computed. These amplitudes can be used along with the reciprocity theorem to compute the wave amplitude from a point-force. The reciprocity theorem used in this case can be explained by considering Fig. 2.11. An incident plane wave can be produced by a force F acting on a remote point in the solid, labelled R, in the direction of the polarization vector of the incident plane wave. This force causes a normal particle displacement on the free surface. The amplitude of this normal displacement on the free surface is given by the superposition of the incident and reflected waves:

$$u_N = A^I p_N^I + A^{qL} p_N^{qL} + A^{qS_V V} p_N^{qS_V} + A^{qS_H} p_N^{qS_H} \tag{2.85}$$

where the subscript N indicates the normal direction. From the reciprocity theorem, a normal force acting on the free surface will cause a particle displacement at R with a magnitude described by the same equation. Therefore all that is required now is a relationship between the force in the solid and the amplitude of the incident wave. This can be computed using the anisotropic Green's function.

2.3.1.2 Green's Function for Anisotropic Media

In an infinite solid the point-force solution to the wave equation is referred to as the Green's function. The displacement u_i at a distance x and time t due to a point-

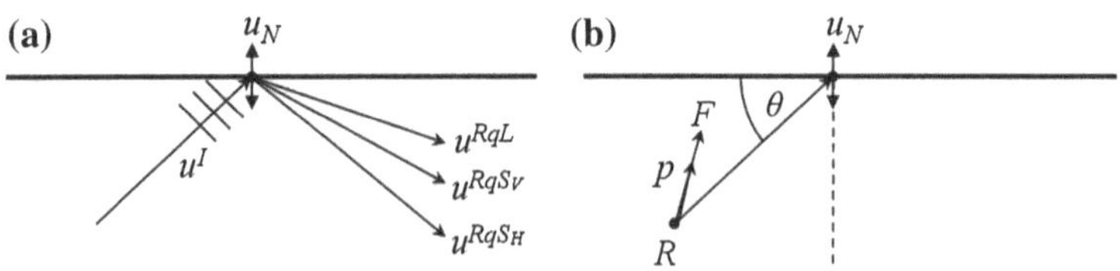

Fig. 2.11 Diagram showing reciprocity theorem that is used to calculate the directivity pattern

force at x_0 and t_0 orientated in the nth Cartesian axis is given by the Green's function G:

$$u_i(x,t) = G_{in}(x,\ t;\ x_0 t_0) \tag{2.86}$$

The time dependence of the force is assumed to be given by a spike, or delta, function. To compute the Green's function for an anisotropic solid, the derivation given by Buchwald [11] using the asymptotic method due to Lighthill [12] is used. This derivation is repeated briefly by Vavryčuk [24] and in detail in Sect. 2.5.5 of Červený [18].

The starting point for the derivation is to return to Cauchy's equation of motion given by Eq. (2.21) but now to include a body force term F:

$$\rho \ddot{u}_i = C_{ijkl} u_{k,\,jl} + \rho F_i \tag{2.87}$$

The body force is assumed to take a general source function representing a small harmonic source given by:

$$F_i = e^{-i\omega t} \iiint\limits_{-\infty}^{\infty} \overline{F}_i e^{ik_j x_j} dk_1 dk_2 dk_3 \tag{2.88}$$

When this force is replaced by a delta function later, the Green's function will be obtained. Making use of the Fourier transform:

$$\overline{y}(k) = \int\limits_{\mathbb{R}^n} y(x) e^{-ikx} dx \tag{2.89}$$

$$y(x) = \frac{1}{(2\pi)^n} \int\limits_{\mathbb{R}^n} \overline{y}(k) e^{ikx} dk \tag{2.90}$$

the 3-D Fourier transform $\overline{F}$ is:

$$\overline{F_i} = \frac{1}{8\pi^3} e^{-i\omega t} \iiint\limits_{-\infty}^{\infty} F_i e^{-ik_j x_j} dx_1 dx_2 dx_3 \tag{2.91}$$

Displacement can also be expressed as:

$$u_i = e^{-i\omega t} \iiint\limits_{-\infty}^{\infty} \overline{u_i} e^{ik_j x_j} dk_1 dk_2 dk_3 \tag{2.92}$$

$$\overline{u_i} = \frac{1}{8\pi^3} e^{-i\omega t} \iiint\limits_{-\infty}^{\infty} u_i e^{-ik_j x_j} dx_1 dx_2 dx_3 \tag{2.93}$$

Therefore, the appropriate differentiation of the above equation and substitution of the results along with Eq. (2.88) into Eq. (2.87) yields:

$$\rho\omega^2\overline{u}_i = C_{ijkl}k_jk_l\overline{u}_k + \rho\overline{F}_i \tag{2.94}$$

This is a set of three linear equations, the solution to which is given by the inversion:

$$\overline{u}_i = \frac{B_{ki}\overline{F}_i}{\rho\det(\Gamma_{ik})} \tag{2.95}$$

where det denotes the determinant of the matrix, and:

$$\Gamma_{ik} = C_{ijkl}k_jk_l - \rho\omega^2\delta_{ik} \tag{2.96}$$

B_{ik} are the cofactors of Γ_{ik}. Therefore inserting (2.95) into Eq. (2.92) yields:

$$u_i = e^{-i\omega t}\iiint\limits_{-\infty}^{\infty} \frac{B_{ki}\overline{F}_i}{\rho\det(\Gamma_{ik})} e^{ik_jx_j}dk_1dk_2dk_3 \tag{2.97}$$

To arrive at Green's function the general force term is replaced by a spike force at x_0 which is orientated along the x_n axis applied at time t_0:

$$F_i = \delta_{in}\Delta(t - t_0)\Delta(x - x_0) \tag{2.98}$$

where δ_{in} is the Kronecker delta function and Δ is the Dirac delta function [18]. By referring to Eq. (2.91) the following expression is obtained for the spike force in the frequency/wavenumber domain:

$$\overline{F}_i = \frac{1}{8\pi^3}\delta_{in}e^{-ik_0x_0} \tag{2.99}$$

Therefore, using $k_i = \omega s_i$, where s_i is the slowness vector (the reciprocal of the phase velocity) yields:

$$G_{in} = \frac{e^{-i\omega t}}{8\pi^3}\iiint\limits_{-\infty}^{\infty} \frac{B_{ki}\delta_{in}}{\rho\det(\Gamma_{ik})} e^{i(k_jx_j - k_0x_0)}dk_1dk_2dk_3 \tag{2.100}$$

Therefore, the Green's function can be computed from the 3D inverse Fourier transform given above. This can be thought of as a superposition of plane waves at arbitrary propagation directions with arbitrary phase velocity. The above equation can be simplified using some involved mathematics. First the residual theorem, (for example see Sect. 1.9 of [4]) is implemented to reduce the volume integral into a surface integral, which physically represents a superposition of plane waves at arbitrary angles but with a phase velocity given by the Christoffel equation. The Green's function can then be computed asymptotically, meaning that the function is computed in the far-field a long distance away from the source, using the method of stationary phase (see Sect. 7.5 of [4]). Therefore, it can be shown that the asymptotic Green's function in an anisotropic material at distance d is given by [18]:

$$G(d)_{in} \approx \frac{p_i p_n}{4\pi\rho\sqrt{|K|}Vd} e^{i\left(kd-\omega t+\frac{\pi}{2}\sigma_0\right)} \tag{2.101}$$

where:

$$\sigma_0 = 1 - \frac{1}{2}sgn(\kappa_1) - \frac{1}{2}sgn(\kappa_2) \tag{2.102}$$

κ_1 and κ_2 are the principle curvatures of the slowness surface and sgn denotes the sign. $K = \kappa_1\kappa_2$ and is referred to as the Gaussian curvature. A method to compute the Gaussian curvature is given in Appendix C. The quantities that vary with direction in the above equation: group velocity, wave vector, and curvature, must be evaluated in the direction of interest.

The anisotropic Green's function shows that the amplitude response from a point-force decreases linearly with distance from the source and is proportional to the reciprocal of the square root of the slowness surface curvature. As this expression contains the reciprocal of the slowness surface curvature, this expression cannot be applied to surfaces with zero curvature, such as the cuspidal edges shown at point B in Fig. 2.7. However, for the qL wavesurface in CMSX-4, the equation is valid at all points.

Therefore, by using Eq. (2.101), the wave amplitude due to a force in an anisotropic solid can be computed. This wave amplitude can be inserted into Eq. (2.84) to find the wave amplitude of the reflected waves. The incident and reflected wave amplitudes are finally used in Eq. (2.85) to compute the wave amplitude due to a normal point-force on the surface of an anisotropic half-space.

2.3.2 Results and Validation

The analytical model was implemented in MATLAB and applied to CMSX-4. The analytical model was used to compute the qL wave mode, as this is the only wave mode that will be used for imaging in subsequent chapters. The computational scheme was as follows. Initially, the elastic tensor was transformed using the Bond matrices to align the crystal axes with the axes of the model using Eq. (2.57). Incident qL plane waves were simulated for angles $-90°$ to $+90°$ relative to the normal to the surface, in $1°$ increments. At each direction the slowness vector describing the incident plane wave was split into its s_1 and s_3 components. The slowness vectors were found for the three reflected wave modes by solving Eq. (2.63) and selecting the results that corresponds to the back reflected waves. The polarization vector of each reflected wave mode was found using (2.62). The slowness and polarization vectors were then used to compute the reflection coefficients using (2.84) which gave the beam directivity due to the free surface in Eq. (2.85). This equation was then multiplied by the anisotropic Green's function, Eq. (2.101), to compute the amplitude of the initial incident plane wave to then compute the final wave amplitude at each direction due to a normal point-force on the CMSX-4 surface. The amplitude was stored as a function of ray direction rather than wavefront normal directions.

As a first check, the model described above was used to compute the beam amplitudes of the longitudinal wave mode from a normal point-force acting on the surface of a isotropic half-space and this was compared to close-form solutions as detailed in Appendix B. The two models were found to be numerically equal to six significant figures.

A numerical finite element (FE) model was developed to validate the analytical model. FE analysis works by dividing a complex model into small, simple elements formed from interconnected nodes. The true displacement field within the elements is approximated by simple functions, referred to as shape functions. Equations are derived to relate the element node forces and displacement based on these shape functions. The equations of the individual elements are assembled in global matrices that describe the behaviour of the entire model. For the FE models described in this thesis, explicit time-marching methods were used to simulate the displacements in the mesh due to a dynamic loading by computing the response at small time-increments. Provided the model elements are sufficiently small, as is the time step over which the calculations are performed, the model will be highly accurate and contain all the wave modes that can be generated. However, due to the number of elements required to obtain a result with sufficient accuracy, FE models demand high computer memory and tend to be much slower than the equivalent analytical solutions. The accuracy of FE models for computing beam amplitude in isotropic media is discussed in Appendix B.

The FE model was developed in the commercial FE modelling software package ABAQUS 6.6 (SIMULIA, Providence RI, USA). The model is shown schematically in Fig. 2.12. The model consisted of three-dimensional cuboid elements formed from eight nodes making up a grid that was 4 nominal wavelengths long by 4 wavelengths deep and 1 wavelength thick, using 100 elements per wavelength. Here the nominal wavelength refers to the wavelength of the slowest qL group velocity as predicted by the analytical model. To reduce the model size, symmetric boundary conditions were used about the $x - z$ and $y - z$ planes at the origin. This was achieved by forcing the displacements in the x direction to zero for nodes that were located at $x = 0$. Likewise y displacements for nodes at $y = 0$ were forced to zero. A point-force was applied to a single node in the model to produce a one cycle pulse of 10 MHz centre frequency. Monitoring nodes were selected that were nominally at 3 wavelengths from the point-force at an angle of 0 to 90° in 1° increments. Due to the discretisation of the FE mesh the nodes were not generally at these nominal positions and so the actual distances and angles were computed and stored. The monitoring nodes had to be placed sufficiently far from the point-force to ensure the three wave modes were time-resolved and could therefore be separated. The Cartesian coordinate axes of the FE model

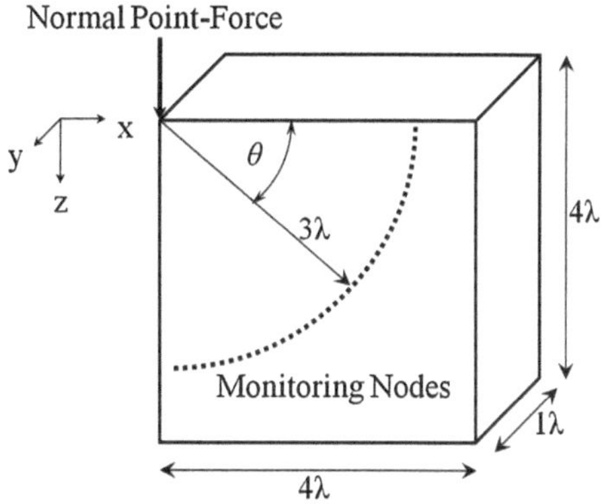

Fig. 2.12 Schematic diagram showing the finite element wave amplitude model

were aligned with the < 001 > crystallographic directions and the monitoring nodes were placed in the {001} plane. The model was run in the time-domain using a time step of 0.5 ns.

Figure 2.13 shows the results of these models computed for a point-force acting normal to the surface with monitoring nodes in the {001} plane at an angle of 0 to 180° in incremental step of 1° with the x, y and z model axes aligned with the < 001 > crystallographic directions. Figure 2.13c shows the absolute wave amplitude from the normal point-force whereas Fig. 2.13a and b show the components of the wave amplitude in the transverse directions (x) and the normal direction (z) respectively. The amplitudes have been normalised to the maximum amplitude value. The two models predict that the amplitude in the remaining direction, that is the normal to the {001} plane (y), is zero. Both models show excellent agreement apart from a small discrepancy around 90° with a maximum error of 10 %. It is thought that the error comes from the fact that the Green's function used in the analytical model is asymptotic. Therefore, it is only exact at very large distances between the source and receiver, whereas, the FE model used a distance of 3 wavelengths to compute the beam amplitude. The models show that the maximum ultrasonic energy in the {001} plane is in the < 011 > direction, which is the direction that has the largest group velocity in this plane. The main discrepancy between the two models is in the direction along the surface where the analytical model predicts zero wave amplitude but the FE model result is non-zero. This is due to the surface waves in the FE model which are not accounted for in the analytical model.

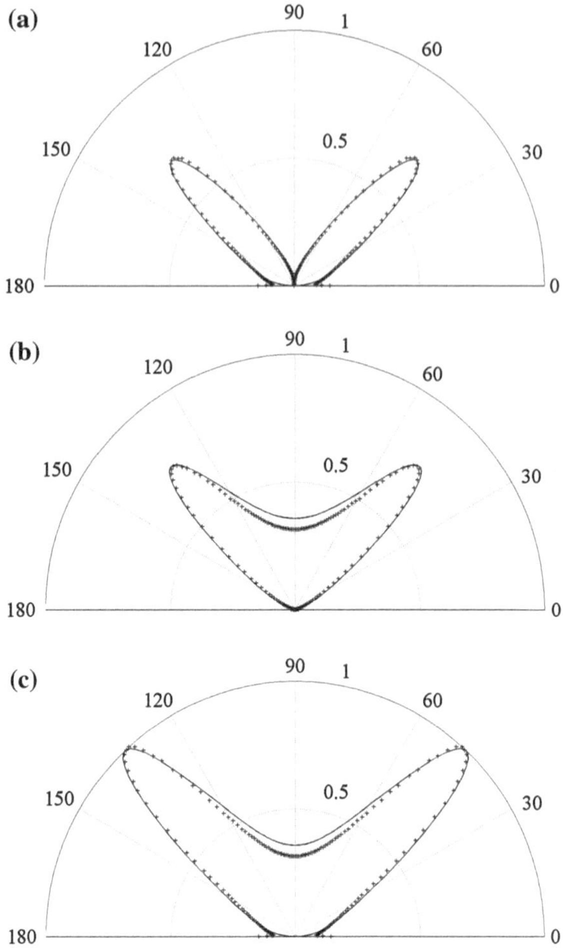

Fig. 2.13 Polar plots showing the beam directivity in the {001} plane with the graphs axes aligned with the < 001 > crystallographic direction from a normal point-force. The plots shows the amplitude in (**a**) the x-direction, (**b**) the z-direction, and (**c**) as an absolute value found using the analytical model (*solid line*) and the FE model (crosses). The amplitudes have been normalised to the maximum amplitude value in this plane

2.4 Summary

This chapter has provided an introduction to the propagation of bulk waves in anisotropic solids. Analytical models have been developed to compute the velocity profiles and wave amplitudes of the three bulk wave modes in anisotropic solids. The models have been applied to CMSX-4 (a single crystal nickel-based superalloy) and the models have been validated using results from numerical finite element models. The analytical models developed will be used in the next chapter. to correct ultrasonic imaging algorithms to account for wave propagation in anisotropic solids

References

1. Love AEH (1944) A treatise on the mathematical theory of elasticity, 4th edn. Dover Publications, New York
2. Green G (1839) On the propagation of light in crystallized media. Trans Cambridge Philos Soc 7(2):121–140
3. Christoffel EB (1877) Ueber die Fortpflanzung von Stössen durch elastische fest Körper. Annali di Matematic 8:193–243
4. Achenbach JD (1975) Wave propagation in elastic solids. Elsevier Science Publishers B. V, Amsterdam
5. Musgrave MJP (1970) Crystal acoustics. Holden-Day, London
6. Kelvin L (1904) Baltimore lectures on molecular dynamics and the wave theory of light. C. J. Clay and Sons, London
7. Musgrave MJP (1954) On the propagation of elastic waves in aeolotropic media. I. general principles. Proc R Soc A 226:339–355
8. Synge JL (1956) Elastic waves in anisotropic media. J Math Phys 35:323–334
9. Merkulov LG (1963) Ultrasonic waves in crystals. Appl Mater Res 2:231–240
10. Fedorov FI (1968) Theory of elastic waves in crystals. Plenum Press, New York
11. Buchwald VT (1959) Elastic waves in anisotropic media. In: Proceedings of the royal society of London. series A, Mathematical and physical sciences. pp 563–580
12. Lighthill MJ (1960) Studies on Magneto-Hydrodynamic waves and other anisotropic wave motions. Philos Trans R Soc A 252:397–430
13. Burridge R (1971) Lamb's problem for an anisotropic half-space. Q J Mech Appl Math 24:81–98
14. Mourad A, Deschamps M (1995) Lamb's problem for an anisotropic half-space studied by the Cagniard de Hoop method. J Acoust Soc Am 97(5):3194–3197
15. Wu K, Nagy PB, Adler L (1990) Far field radiation of a point source on the free surface of semi-infinite anisotropic solids. Rev Prog Quant Nondestr Eval 9:149–156
16. Wu K, Nagy PB, Adler L (1991) Far-field radiation of a vibrating point source in anisotropic media. J Nondestr Eval 10(2):71–78
17. Auld BA (1973) Acoustic fields and waves in solids. Vol 1 and 2, Wiley, New York
18. Červený V (2001) Seismic ray theory. Cambridge University Press, Cambridge
19. Farnell GW (1978) Types and properties of surface waves. In: Oliner AA (ed) Topics in applied physics, Springer-Verlag, Berlin, p 13–59
20. Rose JL (1999) Ultrasonic waves in solid media. Cambridge University Press, Cambridge
21. Miller GF, Musgrave MJP (1956) On the propagation of elastic waves in aeolotropic media. III. Media of cubic symmetry. Proc R Soc A 236:352–383
22. Henneke EG (1972) Reflection-refraction of a stress wave at a plane boundary between anisotropic media. J Acoust Soc Am 51(1):210–217
23. Rokhlin SI, Bolland TK, Adler L (1986) Reflection and refraction of elastic waves on a plane interface between two generally anisotropic media. J Acoust Soc Am 74(4):906–918
24. Vavrycuk V (2007) Asymptotic Green's function in homogeneous anisotropic viscoelastic media. Proc R Soc A 463:2689–2707

Chapter 3
Imaging Anisotropic Components with Ultrasonic Arrays

Abstract An ultrasonic array NDE method was deemed most appropriate for the in situ inspection of the single-crystal turbine blade. Ultrasonic array probes consist of many individually wired ultrasonic elements within a single transducer package. The elements can be fired individually or in groups with different timings or *phase shifts*. This allows an array probe to steer and focus the ultrasonic beam in a way that would otherwise require a great number of different single element transducers. Furthermore, array probes enable ultrasonic imaging of a component from a single inspection location. Ultrasonic imaging involves converting raw ultrasonic data collected by the elements in an array into an image that attempts to accurately represent the internal structure of a component. This allows ultrasonic data to be analysed in a more intuitive manner and reduces interpretation errors. The overall aim of imaging is to improve the probability that a defect is detected by an inspector whilst minimising the chance that a design feature is not mistaken for a defect which could lead to the component being needlessly withdrawn from operation. In this section a novel ultrasonic array imaging algorithm known as the Total Focusing Method (TFM) is introduced. The TFM algorithm is shown to offer many benefits over conventional imaging methods. The algorithm works by post-processing the ultrasonic data, which has been collected using a procedure known as full-matrix capture (FMC), to synthetically focus ultrasonic energy at every point within the image. The FMC process involves transmitting an ultrasonic wave from one element whilst receiving on all other elements in the array. This process is then repeated until the data from every transmit-receive element combination has been collected. The anisotropic behaviour of single-crystal components is shown to significantly distort the images produced with the TFM algorithm. If uncorrected, this distortion could lead to missed defects, false-calls and ultimately an unreliable inspection. Therefore, the TFM algorithm is adapted for imaging anisotropic materials using the models developed in Chap. 2. Simulated and experimental imaging results are presented for both linear and 2D ultrasonic arrays for isotropic and anisotropic (single-crystal) components. The effect of beam profile variation due to the orientation of a single-crystal and the effect of the misalignment of the crystal on imaging performance are demonstrated.

C. Lane, *The Development of a 2D Ultrasonic Array Inspection for Single Crystal Turbine Blades*, Springer Theses, DOI: 10.1007/978-3-319-02517-9_3,

3.1 Introduction

Analytical models that describe the propagation of waves in anisotropic media were developed in the previous chapter. In this chapter, these models are used to correct an algorithm for imaging anisotropic components using ultrasonic phased arrays.

3.1.1 Ultrasonic Array Imaging Algorithms

The algorithm that will be considered is the Total Focusing Method (TFM) [1]. The TFM post-processes the full-matrix capture (FMC) data in the time-domain to synthetically focus ultrasonic energy at every pixel in the image field. The FMC data is collected by pulsing on one element in the array and receiving on all others. This is repeated for every transmit-receive combination. This is a different process to current common practice in industry [2, 3]. Typically the elements are fired together but with time-delays to focus ultrasonic energy at a single depth in the material. The time-delays are then altered to sweep the focused beam. At each sweep angle the scattered wave fields received by the array elements are summed together to produce a sector-scan image. Alternatively, another common method is to use sub-apertures of the array to focus the beam at a single depth. Different sub-apertures are then used to scan the beam along the length of the array. Again, the signals received by the elements in each sub-aperture are summed, and the resulting signals from the different sub-apertures are stacked together to create an image, referred to as a B-scan. The TFM algorithm uses the whole aperture of the array and focuses ultrasonic energy at every depth. As such, the algorithm offers better resolution and defect sensitivity throughout the image compared to the commonly used sector-scan and B-scan methods [1, 4].

There are a number of alternative algorithms that also synthetically focus ultrasonic energy at every location in the imaging field using a linear superposition of the FMC data. These include the back-propagation method [5] and the wave-number algorithm [6]. However, Velichko et al. have proved that these algorithms offer identical imaging performance but effectively use different spatial filters [7]. The TFM algorithm will be used for this thesis as it is the most robust method. The TFM algorithm is also in a form that can be efficiently computed using parallel-processing to allow for a rapid image refresh rate.

There are other algorithms available, which process the FMC data in a fundamentally different way to the linear superposition methods detailed above, that can offer super-resolution imaging [8, 9]. Super-resolution is imaging objects beyond the Rayleigh resolution criterion, and can therefore resolve reflectors that are separated by less than half a wavelength. It is not currently clear how appropriate these types of algorithms are for imaging anisotropic media, and therefore will not be considered any further in this thesis. However, a benefit of collecting the FMC data is that as new algorithms are developed they can be used to process the stored data without having to repeat the inspection.

3.1.2 The Ultrasonic Inspection of Anisotropic Materials

The ultrasonic inspection of anisotropic materials for NDE began to attract a lot of interest in the 1970s. This was primarily due to the need to inspect austenitic steel welds used in the energy and civil-nuclear industries [10]. The grain structure of austenitic welds often exhibits long grains with adjacent grains being crystallographically aligned due to the welding process. Other steel structures tend to have randomly orientated grains with sizes much smaller than typical ultrasonic wavelengths and therefore can be considered isotropic. In the 1980s, Ogilvy produced a computerized model for the propagation of ultrasound through anisotropic austenitic steel welds to aid the development and interpretation of NDE inspections [11, 12].

Directional fibre-reinforced composites also exhibit elastic anisotropy due to differences in elastic stiffness in the direction of the fibres and in the direction normal to the fibres. In 1999, Kline was the first person to correct the synthetic aperture focusing technique (SAFT) algorithm for the variation in velocity in anisotropic fibre-reinforced composites [13]. SAFT differs from the TFM only in that SAFT uses pulse-echo data from individual elements whereas TFM uses FMC data. Consequently TFM tends to offer an improved signal to noise ratio over SAFT, whilst SAFT requires less computational effort.

Spies et al. furthered this work to account for velocity variation in composites [14], and also for the beam profile, using a method Spies had developed based on point-force synthesis [15]. In this work the author scanned a single element probe along the surface of the composite component to collect the pulse-echo data. In 2007, Shlivinski et al. applied the corrected SAFT alogorthim to inhomogeneous anisotropic welds, again using a scanning pulse-echo probe [16]. Later Connolly et al. compared corrected SAFT and TFM images for anisotropic welds using linear ultrasonic arrays [17, 18]. Finally, Chatillon et al. have described commercially available ray-tracing software, CIVA, that can simulate wave propagation from linear and matrix arrays, through components with complex interfaces and through anisotropic media [19].

A number of authors [14, 16, 18] have successfully corrected ultrasonic imaging algorithms to account for the variation in velocity and the beam profiles in various anisotropic materials. In this chapter, the TFM algorithm is corrected for imaging single crystal components using linear and 2D arrays. The effects of velocity variation and beam profile are investigated separately and results are shown for simulated and experimental data. At the end of this chapter, the effect of crystallographic misorientation on the TFM images is examined.

3.2 Anisotropic TFM Algorithm

Consider the volumetric imaging geometry shown in Fig. 3.1. For imaging with a linear array, it will be assumed that the array elements are all located at $y = 0$ and therefore the imaging is constrained to the x–z plane. The TFM algorithm

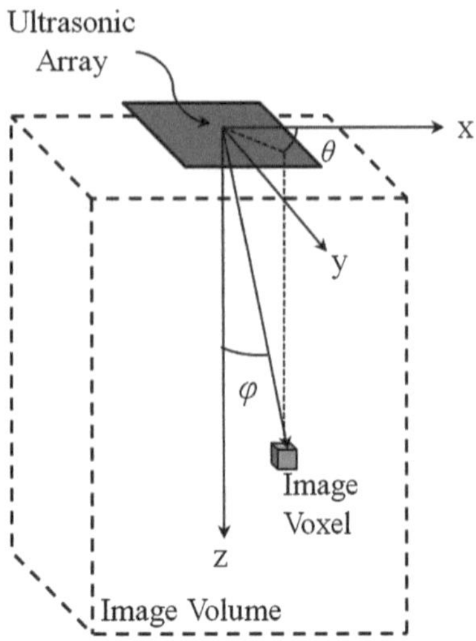

Fig. 3.1 Schematic diagram of volumetric imaging using an ultrasonic array

computes the intensity of a voxel in the imaging field, $I(x,y,z)$, by coherently summing the FMC data, U_{ij}, with appropriate time delays to focus the ultrasonic energy at that location. This can be represented mathematically as:

$$I(x, y, z) = \sum_{i=1,j=1}^{n} A_{ij}(x, y, z)U_{ij}\left(t_{ij}(x, y, z)\right) \tag{3.1}$$

where A_{ij} is an amplitude correction that accounts for the directivity of the individual elements and the beam spread of the ultrasonic waves, n is the number of array elements, and subscripts i and j indicate the transmit and receive elements respectively. Note that tensor notation is no longer being used.

The images obtained from the TFM (or indeed any other linear imaging algorithm) arise from constructive and destructive interference when the signals from different element pairs are combined. For this reason, it might be expected that the phase velocity rather than the group velocity should be used in the modified TFM algorithm for an anisotropic medium. However, this is incorrect. The TFM is based on aligning the travel times of signals that have potentially been scattered from each image point. The travel time between two points in space is determined by the group velocity and therefore it is the group velocity that must be used in the TFM [14]. The constructive and destructive interference necessary to form an image where scatterers are localised at their correct spatial positions then happens quite naturally. The correct time-of-flight, t, of the ultrasonic pulse from the transmit element to the field point and back to the receive element is given by:

$$t_{ij}(x, y, z) = \frac{d_i(x, y, z)}{V(\theta_i(x, y, z), \varphi_i(x, y, z))} + \frac{d_j(x, y, z)}{V\left(\theta_j(x, y, z), \varphi_j(x, y, z)\right)} \tag{3.2}$$

where V is the group velocity defined using spherical coordinates θ and φ, and $d_{i,}$ and $d_{j,}$ are the ray path distances between the image voxel and the transmit and receive elements respectively. These distances are computed using simple trigonometry by:

$$d_i(x,y,z) = \sqrt{\left(x - x_{ij}\right)^2 + \left(y - y_i\right)^2 + z^2} \tag{3.3}$$

$$d_j(x,y,z) = \sqrt{\left(x - x_j\right)^2 + \left(y - y_j\right)^2 + z^2} \tag{3.4}$$

The variable A_{ij} is given by:

$$A_{ij} = \frac{d_i^{1/\mathrm{m}} d_j^{1/\mathrm{m}}}{D_i(\theta_i, \varphi_i, \omega) D_j(\theta_j, \varphi_j, \omega)} \tag{3.5}$$

where D is the directivity of the individual elements. In 2D arrays, waves spread out from elements in three dimensions. The intensity of the wave decreases proportionally to the square of the propagation distance. The acoustic pressure is proportional to the square root of intensity, and thus decreases proportionally with distance. Therefore $m = 1$ for 2D array elements. For linear arrays, the individual elements are considered to be infinitely long in the y-direction. As such, the beam spread is assumed to be restricted to the plane of the array (x–z plane). The intensity thus decreases at a rate proportional to the propagation distance, and therefore $m = 2$ for linear arrays.

3.3 Effect of Velocity Variation

The effect of velocity variation on the TFM algorithm will now be investigated in this section. The effect will be examined using both simulated and experimental array data. The effect due to beam directivity is considered separately in the next section. As such, the variable A is set to equal unity in the forward model used to compute the simulated data and in the TFM algorithm. Two ultrasonic phased array probes will be considered: a linear array and a 2D array. The specifications for the two arrays are given in Table 3.1. The element pitch refers to the distance between adjacent elements. Both arrays were manufactured by Imasonic (Besançon, France). The elements of both probes transmit ultrasonic energy predominantly in

Table 3.1 Properties of the two ultrasonic arrays

Array	Number of elements	Layout	Element pitch (mm)	Centre frequency (MHz)	Wavelength at 6000 m/s (mm)
Linear	64	Linear (64 × 1)	0.21	15	0.4
2D	121	Matrix (11 × 11)	1.0	3	2

the longitudinal mode. In addition, the longitudinal mode is the quickest, and as such, the scattered longitudinal waves tend to be well separated from other reflected modes when received by the array elements. Therefore, throughout this thesis imaging with be performed using the (quasi-)longitudinal mode.

3.3.1 Simulated Data

The following forward model was used to simulate array data for imaging a point-like reflector using a script written in MATLAB. Initially, a five cycle wave packet with the appropriate centre frequency was created. Time shifts were then applied to the wave packet to simulate the wave propagating from the transmit element to the point-like reflector, and then from the reflector back to the receive element. This was done for every possible transmit-receive combination to simulate the FMC data. Data was simulated in this way for both arrays and also for both isotropic media (with a constant group velocity of 6060 m/s) and anisotropic media. The anisotropic simulations used the quasi-longitudinal group velocity profile of single crystal CMSX-4 computed in the previous chapter with the $< 001 >$ crystallographic directions aligned with the imaging axes. For the linear array the point reflector was simulated under the centre of the array at a depth of $z = 7.5$ mm. The point reflector depth was at $z = 10$ mm for the 2D array simulation.

The TFM algorithm was run on all sets of data using MATLAB. In the anisotropic case, the data was first processed with constant velocity of 6060 m/s and then the corrected algorithm was implemented using the correct anisotropic velocity profile.

Figure 3.2 shows the TFM images for the simulated linear array. The images have been displayed in decibels (dB) which can be computed as:

$$dB(x, y, z) = 20 \log_{10} \frac{|I(x, y, z)|}{\max(|I|)} \tag{3.6}$$

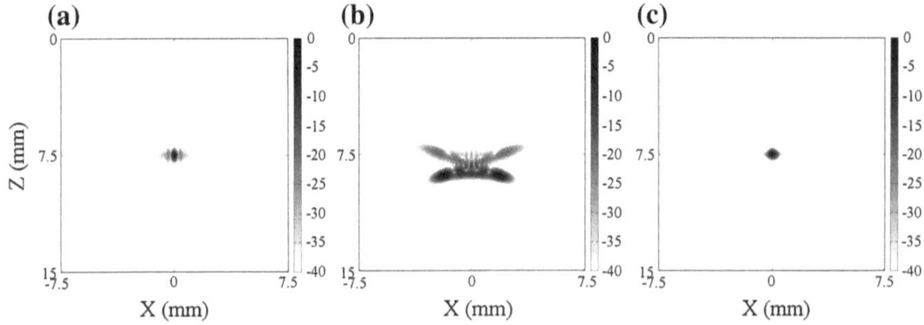

Fig. 3.2 TFM images from simulated data for the linear array for the **a** isotropic, **b** uncorrected anisotropic and **c** corrected anisotropic case. The images are displayed using the dB scale

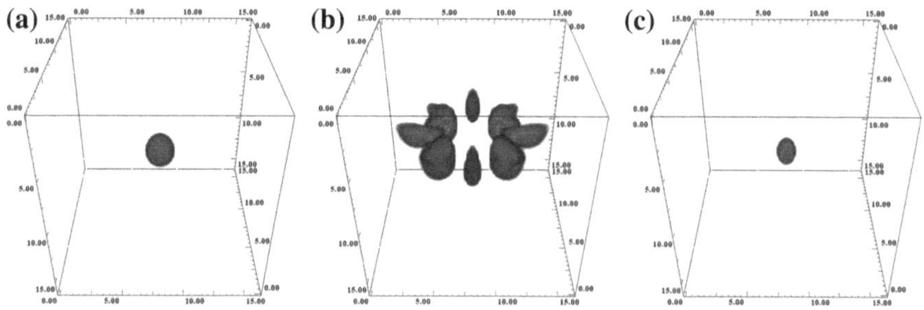

Fig. 3.3 Volumetric TFM images from simulated data for the 2D array for the **a** isotropic, **b** uncorrected anisotropic and **c** corrected anisotropic case. The images are displayed by rendering all isosurfaces 12 dB below the peak signal. Dimensions are shown in units of mm

Figure 3.2a shows the isotropic TFM image of the point-like reflector. The reflector is displayed as a single point at the correct location at $x = 0$ and $z = 7.5$ mm. Figure 3.2b shows the TFM image from the anisotropic single crystal specimen that was computed assuming constant velocity in all directions. The image of the point-like reflector can be seen to be significantly distorted. When the correct velocity profile for the simulated single crystal data was used, as shown in Fig. 3.2c, the TFM image is refocused and closely resembles the isotropic image.

The TFM images of the simulated isotropic and anisotropic materials for the 2D array are shown in Fig. 3.3. The volumetric images were computed using MATLAB but were displayed using VGStudio Max (Volume Graphics GmbH, Heidelberg, Germany). The images were created by rendering all isosurfaces (connected surfaces of equal value) 12 dB below the peak signal. The TFM image of the point reflector in the isotropic material, Fig. 3.3a, shows a focused single point in the correct location. In the uncorrected anisotropic case, shown in Fig. 3.3b, the point reflector is defocused to a great extent with eight points imaged at incorrect locations. However, when the correct anisotropic profile is used, as shown in Fig. 3.3c, the image focus is restored and the morphology of the indication is close to the isotropic case.

3.3.2 Experimental Data

The corrected TFM algorithm was then used on actual experimental data. Figure 3.4 shows diagrams of the linear and 2D array test specimens used. The linear array test specimen was a 15 × 15 × 30 mm block with a 0.125 mm wide slot wire-cut from one face to a depth of 7.5 mm. At this width the tip of the slot approximates a point-like reflector as it is much less than the nominal longitudinal wavelength of 0.4 mm. The linear array test pieces were made from a polycrystalline equiaxed nickel alloy with an isotropic longitudinal velocity of 6060 m/s

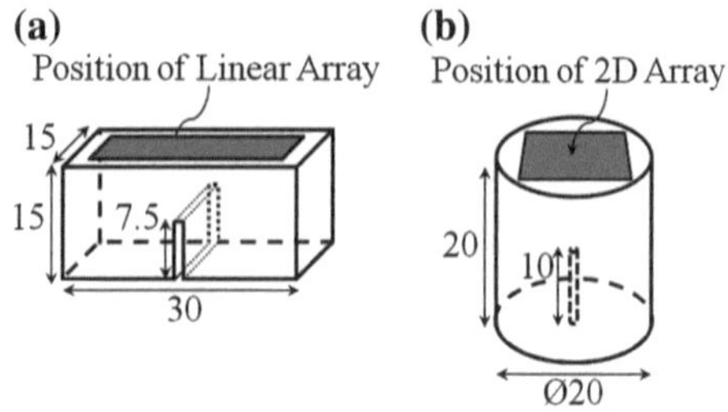

Fig. 3.4 Schematic diagrams of **a** the linear array test piece and **b** the 2D array test piece. Dimensions are in mm

and an anisotropic single crystal CMSX-4. The axes of the single crystal test block were approximately aligned with < 001 > crystallographic directions.

Figure 3.4b shows the cylindrical test piece design used for the 2D array experiments. The test pieces contained a 1 mm diameter spherical-ended hole drilled from the centre of one of the flat surfaces. Again, the two test pieces were manufactured from an isotropic nickel alloy and anisotropic CMSX-4. The single crystal test specimen was cast approximately in the [001] direction. At 3 MHz, the nominal longitudinal wavelength in the specimens was 2 mm. Therefore, the 1 mm diameter spherical-ended hole also approximates a point-like reflector.

The FMC data was collected using the Imasonic array probes described in Table 3.1 and a 128-channel Micropulse MP5PA array controller manufactured by Peak NDT Ltd (Derby, UK). A band-pass filter was applied to the time-traces to reduce incoherent noise. Example time-traces from the linear and 2D array are shown in Fig. 3.5.

The TFM was run on all sets of collected data. In the anisotropic case, the TFM images were processed using a constant velocity of 6060 m/s and the *correct* anisotropic velocity profile. The velocity profiles were found by first measuring actual crystallographic orientation using a Laué X-ray diffraction method [20]. This measuring system is highly accurate, able to measure crystallographic orientation to $\pm$ 0.8°, and is therefore regarded throughout the thesis as providing the true crystallographic orientation. Once the orientation was known, the velocity profiles were rotated in line with the crystallographic axes.

Figure 3.6 shows the TFM images for the linear array experiments. The images are displayed in dB with 0° dB set to the peak backwall signal of each image. Figure 3.6a shows the TFM image of the isotropic test piece. The image shows the backwall, correctly located at z = 15 mm, as a flat line and an indication of the tip of the wire-cut slot centred on 7.5 mm. The TFM image of the single crystal CMSX-4 specimen that assumed a constant group velocity, Fig. 3.6b, exhibits significant defocusing and distortion to the backwall and slot tip features. On the other hand, the corrected image, Fig. 3.6c, is very similar to the isotropic image. This shows that the adapted algorithm was implemented effectively.

Figure 3.7 shows the volumetric TFM images of the 2D array isotropic test piece, and the uncorrected and corrected images for the anisotropic single crystal test piece. VGStudio Max was used to create the images by displaying the iso-surfaces 35 dB below the peak backwall signal in each image. The defocusing

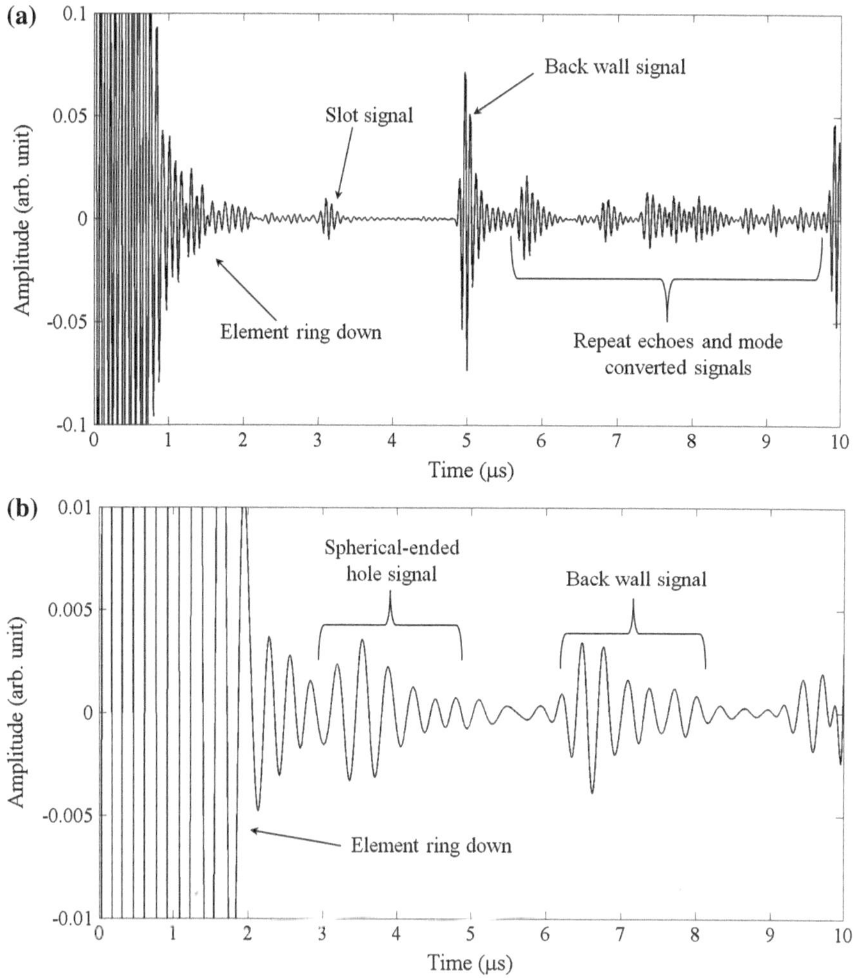

Fig. 3.5 Example time-traces from **a** the 15 MHz linear array and **b** the 3 MHz 2D array. Both examples shown are for transmitting and receiving on the first element in the array

effect of the anisotropy can be seen to cause significant distortion in the uncorrected image. The uncorrected anisotropic image was not sufficiently clear to image the spherical ended hole and the reconstructed image exhibits many image artefacts from inter-element cross talk and element ring down. Once the corrected algorithm is applied, the image is refocused and the backwall and spherical-ended hole indication can be easily indentified.

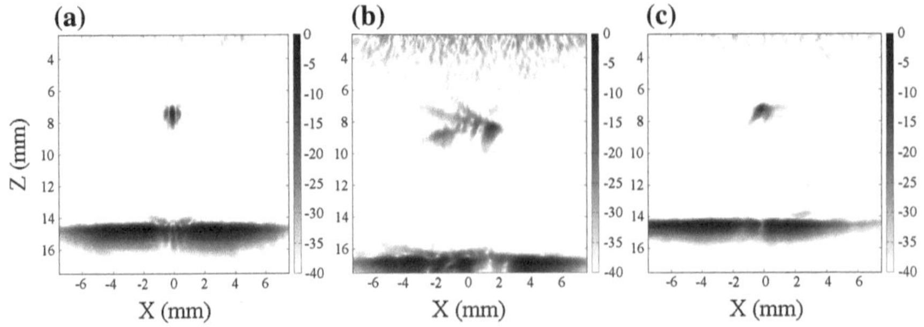

Fig. 3.6 TFM images from experimental data for the linear array for the **a** isotropic, **b** uncorrected anisotropic and **c** corrected anisotropic case. The images are displayed in dB with 0 dB set to the peak *back* wall signal

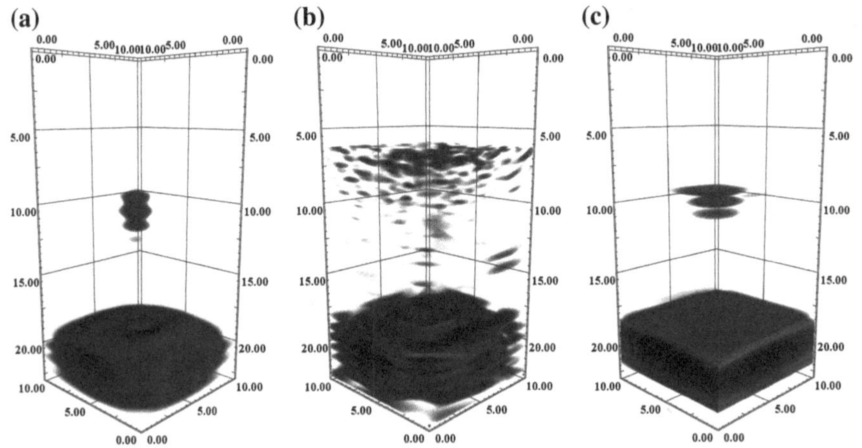

Fig. 3.7 Volumetric TFM images from experimental data for the 2D array for the **a** isotropic, **b** uncorrected anisotropic and **c** corrected anisotropic case. The images are displayed by rendering all isosurfaces 35 dB below the peak *back* wall signal. Dimensions are shown in units of mm

3.4 Effect of Beam Profile Variation

It has been shown in the above section that the anisotropic velocity profile has a significant effect on the TFM image. If isotropic properties are assumed when imaging anisotropic single crystal components, the resulting images can become very distorted. In the case of the 2D array experiment the uncorrected image from the spherical bottomed hole could not be distinguished from the image noise caused by inter-element cross talk and element ring down.

In this section, the effect of beam profile is investigated. It is assumed that the correct velocity profile is now always used when computing the images. It can be seen from the TFM Eq. (3.1) that the effect of beam profile is to vary the amplitude of a point-like reflector within an image. The complex scattering from specular reflectors will further alter the image amplitude of these non-point-like reflectors depending on their location and orientation relative to the array. However, in this study this complicated scattering effect is neglected and only point-like reflectors are considered for the sake of simplicity.

In this section, the variation in beam amplitude from only the linear array will be considered, but the following analysis is equally valid for the 2D array or indeed any arbitrary array.

3.4.1 Simulated Field Patterns

To simulate the field pattern from an array, the change of beam amplitude with direction, or *directivity*, from the individual elements of the array must first be computed. This is achieved using point-force synthesis [15] based on Huygens' principle. Here the element is divided up into point-forces and then the resulting field is created from a summation of the beam from the point-forces. The directivity of a point-force, D_{point}, in CMSX-4 was computed in the previous chapter. Therefore, the element directivity D is given by:

$$D(\theta, \varphi, \omega) = \sum_{i=1}^{p} \frac{e^{-\frac{i\omega d_i}{V(\theta_i, \varphi_i)}}}{d_i} D_{point}(\theta_i, \varphi_i) \tag{3.7}$$

where p is the number of point-forces used.

For linear arrays it is common to model the length of the elements as infinite and, as such, only the variation in beam amplitude as angle φ changes, whilst angle $\theta = 0$, is typically considered. However, for elements in 2D arrays, the variation in beam amplitude in all directions must be modelled.

In isotropic media the group velocity is constant and therefore the summation can be replaced with an integration over the width of the element, a, which can be solved analytically [3] to yield:

$$D(\varphi, \omega) = \text{sinc}\left(\frac{\pi a \sin(\varphi)}{\lambda}\right) D_{point}(\varphi) \tag{3.8}$$

To clarify, the sinc function in the above equation is the unnormalised sinc function. For an isotropic material, the point-force directivity can be found using the same analysis given in Sect. 2.3. However, for simplicity, the widely used analytical equation given by Miller and Pursey [21] was implemented. For a normal point-force acting on the surface of an isotropic half-space, the directivity is given by:

$$D_{point}(\varphi) = \frac{\left((c_l/c_s)^2 - 2\sin^2(\varphi)\right)\cos(\varphi)}{F(\sin(\varphi))} \tag{3.9}$$

where:

$$F(\zeta) = \left(2\zeta^2 - \left(\frac{c_l}{c_s}\right)^2\right)^2 - 4\zeta^2\left(\zeta^2 - 1\right)^{1/2}\left(\zeta^2 - \left(\frac{c_l}{c_s}\right)^2\right)^{1/2} \tag{3.10}$$

Figure 3.8 shows the beam directivity patterns from a 0.2 mm wide 15 MHz array element radiating into isotropic nickel and anisotropic CMSX-4. The directivity patterns for the anisotropic case were computed using 50 point-forces along the width of the element, approximately equal to 100 points per wavelength The point-force directivity for both cases is also shown. It can be seen that the element pattern closely matches that of the point-force for both isotropic and anisotropic material. This is to be expected, as the element width is much smaller than the longitudinal wavelength at 15 MHz, and therefore the element approximates a point-force. The directivity pattern in the isotropic nickel, shown in Fig. 3.8a, shows that the majority of beam energy is directed normal to the element surface. The amplitude then decreases moving away from the normal, and becomes zero in the direction parallel to the surface. The anisotropic case is significantly different to the isotropic case. In the {001} plane shown in Fig. 3.8b,

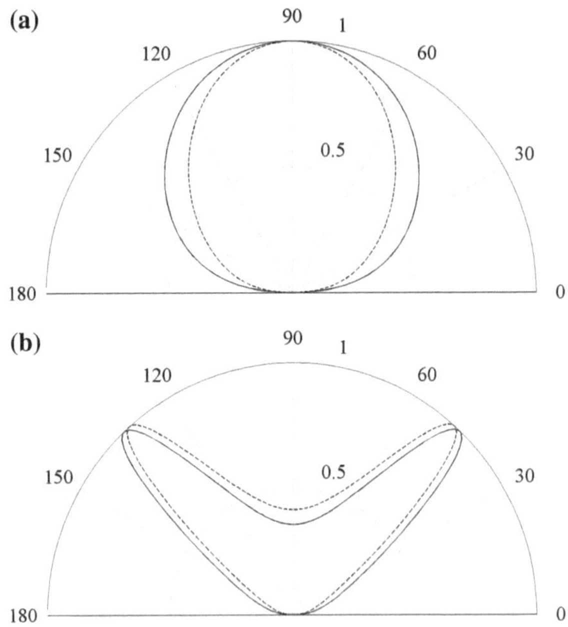

Fig. 3.8 Polar plot showing the beam directivity for a point-force (*solid line*) and a 0.2 mm 15 MHz element (*dashed line*) for **a** isotropic nickel and **b** CMSX-4. The CMSX-4 result is shown for the {001} plane and with the $< 001 >$ crystallographic directions aligned with the Cartesian axes

the majority of beam amplitude is in the $< 011 >$ direction which corresponds to the direction of highest group velocity.

The beam directivities of the individual elements were then used to compute the TFM beam fields in isotropic nickel and anisotropic CMSX-4 using different arrays. These are effectively plots of amplitude response from a point-like reflector at every location in the image using the TFM algorithm. These were computed by summing the beam directivity contributions for every transmit-receive element combination together with the beam spread for every pixel in the TFM image. As the linear array was modelled, the spread of the ultrasonic beam was assumed to be confined to the imaging plane, and therefore the $\sqrt{d}$ relationship was used for beam spread.

Figures 3.9, 3.10, 3.11 and 3.12 show the TFM beam fields for the full 64 element 15 MHz array and for an aperture formed from 4 adjacent elements in the isotropic nickel and anisotropic CMSX-4. In all examples, a plot of field amplitude for a line located at a depth of 7.5 mm running from $x = -10$ to $x = 10$ mm is shown. In the isotropic case, as shown in Figs. 3.9 and 3.10, the amplitude decreases with increasing depth in the material and moving sideways from the

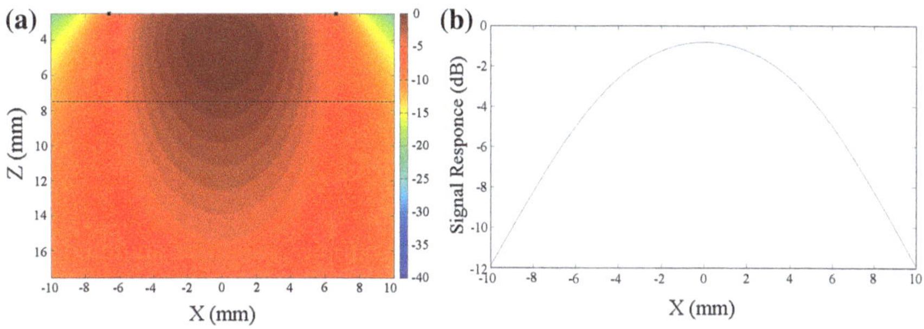

Fig. 3.9 Beam field for the 64 element array in isotropic nickel. The aperture of the array is indicated using *black* crosses on the top of the graph **a**. The signal response shown in **b** is from the position of the *dashed line* shown

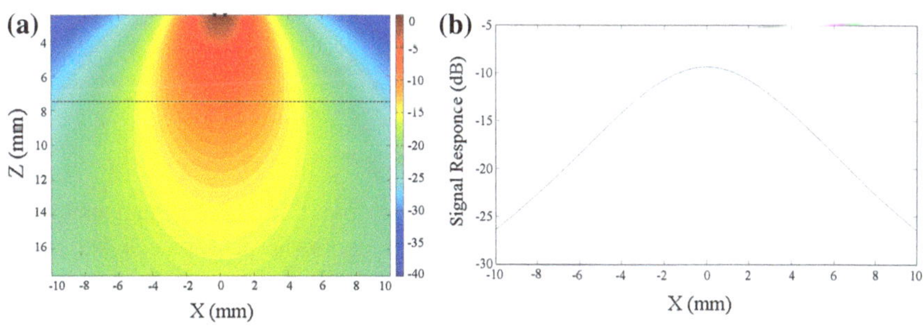

Fig. 3.10 Beam field for the 4 element aperture in isotropic nickel

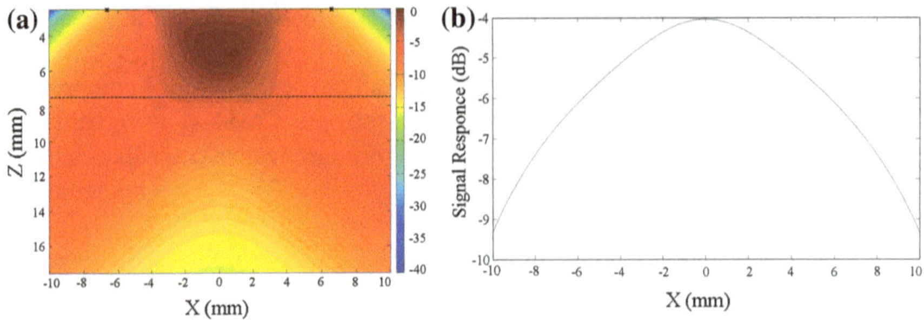

Fig. 3.11 Beam field for the 64 element array in CMSX-4

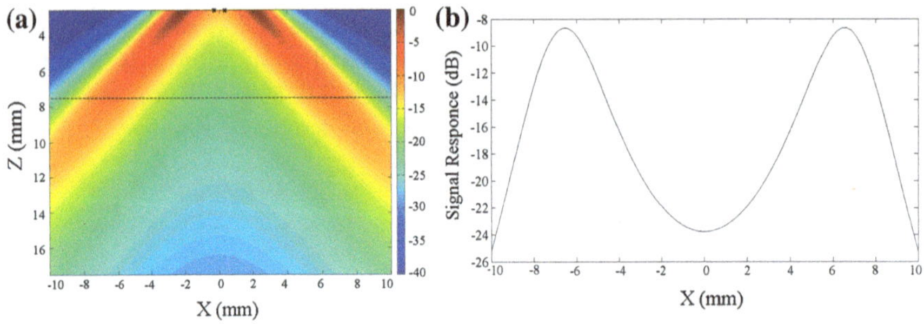

Fig. 3.12 Beam field for the 4 element aperture in CMSX-4

centre of the array. This is not so for the anisotropic case. The amplitude still decreases with depth, along a direction pointing from the centre of the array, due to beam spread. However, at a given depth the field amplitude increases in the $< 011 >$ direction; the direction of highest group velocity for the plane shown. This is shown most markedly in the 4 element aperture in Fig. 3.12. This means reflectors located at a $< 011 >$ crystallographic direction relative to the centre of the array will have a much greater TFM signal than equivalent reflectors located directly beneath the centre of the array for the crystallographic orientation used in this simulation.

3.4.2 Experimental Results

To validate the simulations, the following experiment was performed using the 15 MHz linear array and the isotropic and anisotropic test pieces shown in Fig. 3.4a. The array was centred directly above the tip of the wire-cut slot and the FMC data was captured. The TFM response from the tip of the slot was then computed using a sub-aperture of 4 elements. The sub-aperture was electronically scanned along the length of the 64 element array. This is equivalent to moving the tip of the reflector relative to the centre of the 4 element sub-aperture, but eliminates positioning inaccuracies.

To ensure the results were not affected by the variation in the sensitivity of the individual elements, their response was first measured by recording the amplitude of a pulse-echo reflection from a backwall in an isotropic plate. This process was repeated five times to remove errors due to coupling, and the average response was computed. The signals from the tip of the wire-cut slot were then normalised using this data.

Figure 3.13 shows the TFM amplitude response from the tip of the slot in the isotropic test piece for the different 4-element sub-apertures. The results from the experiment, which was performed three times, show a consistent profile indicating that the coupling was consistent for all experiments. The experimental profile shows fairly good agreement with the predicted result. The main error is thought to arise from the tip of the wire cut slot not perfectly resembling a point-like reflector. Other smaller errors may be due to slight variations in the element performance.

Figure 3.14 shows the TFM amplitude response from the tip of the slot in the anisotropic CMSX-4 test piece for the different 4-element sub-apertures. The profile can be seen to differ substantially from the isotropic profile, with the

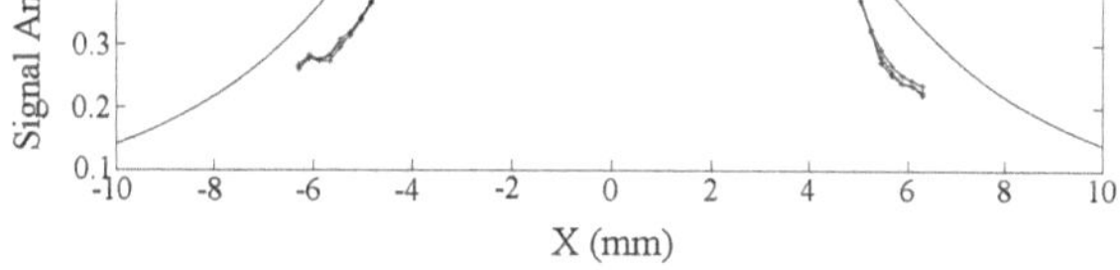

Fig. 3.13 The point-like reflector signal amplitude response in an isotropic medium from the 15 MHz array using a four element aperture for predicted (*solid line*) and three experimental results (*lines with crosses*)

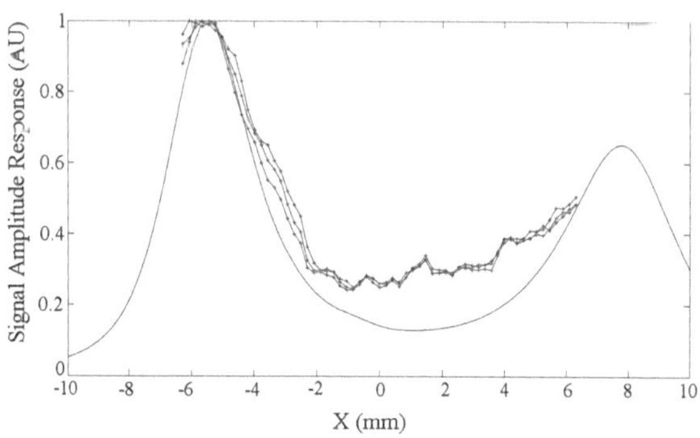

Fig. 3.14 The point-like reflector signal amplitude response in CMSX-4 from the 15 MHz array using a four element aperture for predicted (*solid line*) and three experimental results (*lines with crosses*)

maximum response occurring off-centre in the direction of highest group velocity. The plot is not symmetric due to crystallographic orientation of the component. For the simulated results, the $< 001 >$ crystallographic directions were aligned with the Cartesian axes. However, in the experimental case the [001] direction was misaligned relative to the Z-axis by approximately 5°. The experimental results again show reasonable agreement with the predicted results, with the main error being due to the non-perfect point-like nature of the tip of the wire-cut slot. This conclusion on the source of the error could potentially be further validated by performing the experiments at a lower frequency and hence the wire-cut slot would become more point-like relative to the ultrasonic wavelength.

The experiment just described has shown that the beam amplitude correction will alter the signal from an identical reflector depending on where the reflector is relative to the array. The need to implement the correction for beam profile depends, therefore, on the inspection performed. For example, if an inspection has a rejection criterion based on local SNR the correction will have little effect as it will alter the signal and noise in the same location by the same amount. However, if the rejection is based on the amplitude from a defect, which could occur at any location in the image, being above a set threshold then the beam amplitude correction should be considered.

3.5 Effect of Crystallographic Misorientation

The above analysis has shown that using a constant velocity profile when imaging anisotropic single crystal components causes significant distortion to the TFM image. In this section, the effect on the TFM image of using the incorrect velocity profile caused by a misorientation in the single crystal is investigated. This is to determine how accurately the crystallographic orientation must be measured before a single crystal component can be successfully imaged. In this section, the beam profile is ignored and therefore $A = 1$ for the forward model and the TFM algorithm.

3.5.1 Simulations with the Linear Array

The orientation of a cubic crystal, and hence the velocity profile, can be described by three consecutive rotations around the $< 001 >$ crystallographic directions. To investigate the effect of these rotations, simulated data for imaging with the 15 MHz linear array was used. The rotations were considered separately and the data was simulated for a point-like reflector located at $z = 7.5$ mm, using the velocity profile computed at each rotation. The TFM algorithm was then run on the simulated data, using a velocity profile that assumed the $< 001 >$ crystallographic directions were aligned with the image axes.

Figures 3.15, 3.16 and 3.17 show the TFM images for 5°, 10° and 45° rotations about the z, y and x image axes respectively. It can be seen that the TFM images of the point-like reflector change very little with rotation around the z-axis. A small rotation around the y-axis causes a significant distortion to the reflector indication. A large rotation around the x-axis is required to cause a similar amount of defocusing.

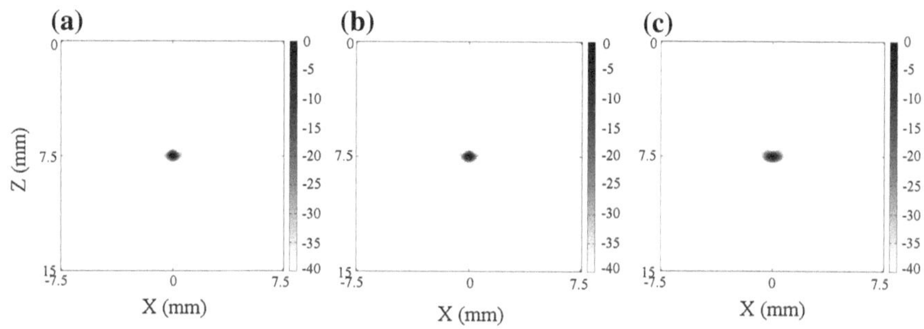

Fig. 3.15 Simulated TFM image of the point-like reflector with **a** 5°, **b** 10° and **c** 45° rotation of the crystal about the image z-axis processed with a velocity profile that assumed no rotation

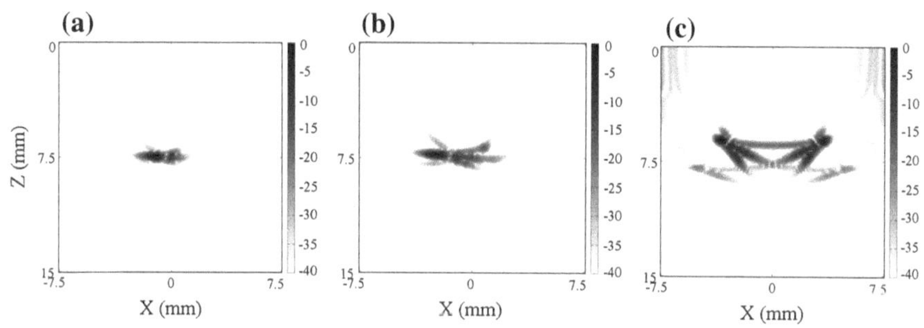

Fig. 3.16 Simulated TFM image of the point-like reflector with **a** 5°, **b** 10° and **c** 45° rotation of the crystal about the image y-axis processed with a velocity profile that assumed no rotation

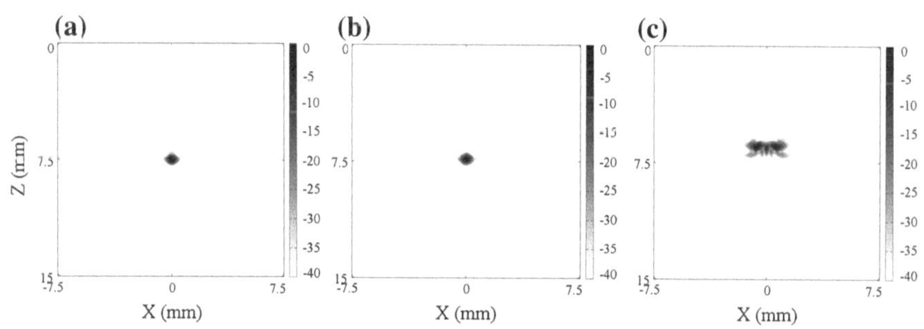

Fig. 3.17 Simulated TFM image of the point-like reflector with **a** 5°, **b** 10° and **c** 45° rotation of the crystal about the image x-axis processed with a velocity profile that assumed no rotation

These effects can be explained by examining the velocity profiles used for the simulations as the individual rotations occur, as shown in Fig. 3.18. On each graph the maximum range of angles between the array elements and the point-like reflector is indicated using vertical dashed lines. This represents the angles that the reflector is imaged over. As Fig. 3.18a shows, negligible change occurs between the imaging angles with a rotation about the z-axis and therefore the TFM image changes negligibly. The velocity profile changes rapidly with rotations about the y-axis, as shown in Fig. 3.18b, hence a small rotation causes a large defocusing of the TFM image. Finally, Fig. 3.18c shows that, for the rotations about the x-axis, a large change only occurs in the velocity profile used for imaging at 45°, and correspondingly the TFM images only become distorted at this rotation.

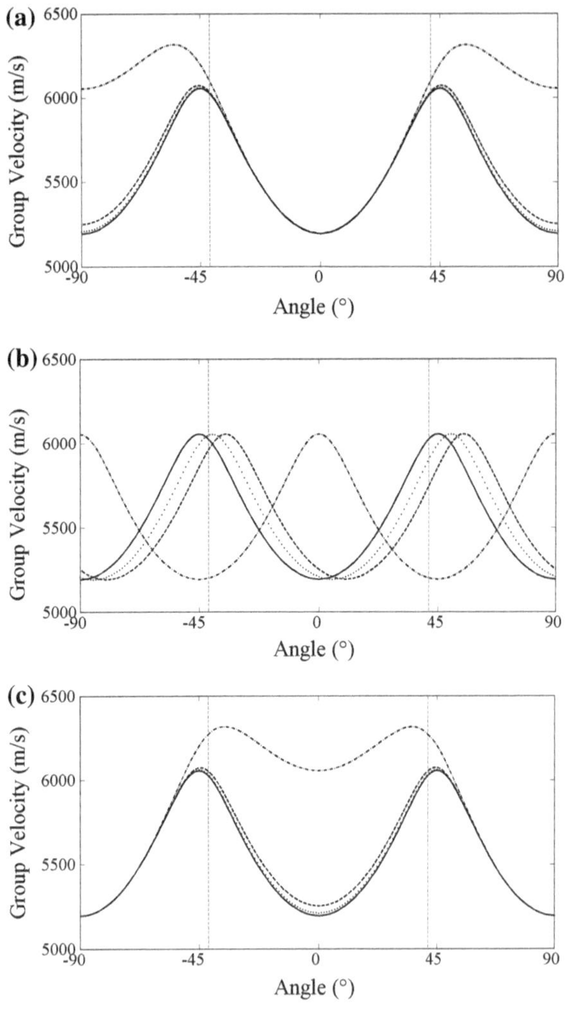

Fig. 3.18 Velocity profiles due to a rotation about the **a** z, **b** y and **c** x axes of 0° (*solid line*), 5° (*dotted line*), 10° (*dashed line*), and 45° (*dash-dot line*)

3.5.2 Experiments with the 2D Array

The analysis with the linear array shows that misorientation in a single crystal can cause significant distortion to the TFM image. However, for NDE detection inspections it is not the morphology of the defect indication that is of importance, but the amplitude of the defect relative to the background image noise. The defect indication morphology is important when it comes to sizing and characterisation. Therefore, the following experiment using the 2D array and the 2D array test piece was performed to investigate the effect of orientation on signal to noise ratio (SNR). This was achieved by imaging the single crystal test piece using the orientation found from the X-ray diffraction method [20]. The crystallographic orientation used in the imaging algorithm was then deviated from the true orientation by rotations around each of the three $< 001 >$ crystallographic directions in turn. After each iteration, the image was re-processed and the SNR was computed.

Using the true orientation the signal amplitude in the TFM image from the spherical-ended hole relative to the peak backwall was -21 dB. This was 17 dB above the maximum level of noise at -38 dB. It was found that as the crystallographic orientation was varied from the true orientation, the hole signal decreased and the level of noise increased, as shown in Fig. 3.19. In addition, the location of the spherical-ended hole indication in the image moved away from its actual position. For successful detection the signals from defects should always be above a threshold which is set to be greater than the maximum level of noise. This investigation showed that for the spherical-bottomed hole signal to be 12 dB greater than the maximum level of noise, the crystallographic orientation used for the imaging algorithm should not deviate from the true orientation by more than $\pm$ 4° in the x and y axes. The rotation around the z axis had negligible effect on the SNR as the z-axis was closely aligned to a $< 001 >$ direction.

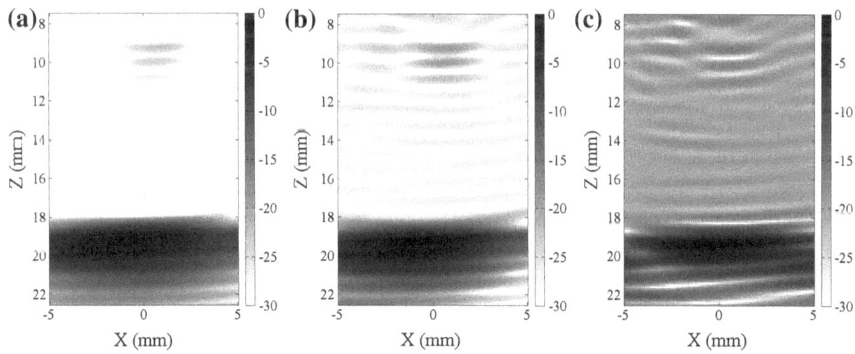

Fig. 3.19 TFM images of the 2D array specimen computed using the velocity profile from a crystallographic orientation that was misorientated in y-axis by **a** 0°, **b** 5° and **c** 10° from the true orientation. The TFM images are shown in the X–Z plane at Y = 0 mm. The images are displayed in dB with 0 dB set to the peak *back* wall signal in image *a*

A similar experiment was performed using the linear array and the linear array test piece. It was found that although the image of the defect became highly distorted, the difference between signal and noise never dropped below 21 dB. This was because the signal to noise ratio was much higher in the collected time-traces for the linear array than the 2D array, as can be seen in Fig. 3.5.

3.5.3 Discussion on Crystallographic Misorientation

It has been shown that the misorientation of a single crystal can significantly affect the TFM image. However, a definitive result that dictates the accuracy of the crystallographic orientation measurement prior to ultrasonic imaging cannot be determined. This is because the effect of misorientation on the inspection reliability depends on many factors. For example, it has been shown that the axis of misorientation is important. In addition, the geometry of the array and the position of reflectors are significant, as these determine the range of angles over which the reflector is insonified. Furthermore, it was claimed above that for defect detection inspections the signal to noise ratio is the critical factor and not the indication morphology. This, however, is not strictly true. For example, a component may have a complex internal structure, and therefore defect signals could get misinterpreted for structural features and vice versa.

Therefore, the conclusion that must be drawn from this analysis is that the misorientation is significant but its effects must be determined on an individual inspection basis. Therefore, when the in situ array inspection for the turbine blades is developed in Chap. 5, a similar analysis will be performed to compute the required orientation accuracy for that specific inspection.

3.6 Summary

The anisotropic nature of single crystal materials can significantly distort the TFM images produced using linear and 2D ultrasonic arrays, reducing the sensitivity of these inspections for non-destructive evaluation. An analytical model of bulk wave propagation in anisotropic media has been used to correct an array imaging algorithm to produce highly focused images. The effectiveness of the corrected algorithm has been demonstrated using simulated and experimental data for both isotropic and single crystal specimens.

The effect of beam profile has also been investigated. It has been shown that the beam amplitude correction will alter the signal from a point-like reflector depending on where the reflector is relative to the array.

An investigation looking at the effect of crystallographic misorientation on the TFM image was presented in the final section of this chapter. It was shown that the misorientation could significantly affect the morphology and SNR of a reflector and should therefore be measured prior to inspection. However, it was concluded that the required accuracy of the orientation measurement depends on the parameters of the inspection being performed and therefore must be computed on an individual inspection basis.

References

1. Holmes C, Drinkwater BW, Wilcox PD (2005) Post-processing of the full matrix of ultrasonic transmit-receive array data for non-destructive evaluation. NDT&E Int 38(8):701–711
2. *Introduction to Phased Array Ultrasonic Technology Applications*. Advanced Practical NDT Series. 2004, Waltham, Massachusetts: R/D Tech Corp
3. Drinkwater BW, Wilcox PD (2006) Ultrasonic arrays for non-destructive evaluation: a review. NDT & E Int 39(7):525–541
4. Drinkwater BW, Bowler AI (2009) Ultrasonic array inspection of the clifton suspension bridge chain-links. Insight–Non-Destr Test Condition Monit 51(9):491–498
5. Velichko A, Wilcox PD (2009) Reversible back-propagation imaging algorithm for postprocessing of ultrasonic array data. IEEE Trans Ultrason Ferroelectr Freq Control 56(11):2492–2503
6. Hunter AJ, Drinkwater BW, Wilcox PD (2008) The wavenumber algorithm for full-matrix imaging using an ultrasonic array. IEEE Trans Ultrason Ferroelectr Freq Control 55(11):2450–2462
7. Velichko A, Wilcox PD (2010) An analytical comparison of ultrasonic array imaging algorithms. J Acoust Soc Am 127(4):2377–2384
8. Simonetti F (2007) Imaging beyond the born approximation: an experimental investigation with an ultrasonic ring array. Phys Rev E 76(3):36601–36611
9. Simonetti F (2009) Modern ultrasound imaging: an exciting window into the sub-surface world. Rev Prog Quant Nondestr Eval 28:33–46
10. Connolly GD (2009) Modelling of the propagation of ultrasound through austenitic steel welds. Ph.D. Thesis, department of mechanical engineering, imperial college London: UK
11. Ogilvy JA (1985) Computerized ultrasonic ray tracing in austenitic steel. NDT Int 18(2):67–77
12. Ogilvy JA (1986) Ultrasonic beam profiles and beam propagation in an austenitic weld using a theoretical ray tracing model. Ultrasonics 24(6):337–347
13. Kline RA (1999) SAFT imaging in anisotropic media. Rev Prog Quant Nondestr Eval 18:913–919
14. Spies M, Jager W (2003) Synthetic aperture focusing for defect reconstruction in anisotropic media. Ultrasonics 41(2):125–131
15. Spies M (1994) Transducer-modeling in general transversely isotropic media via point-source-synthesis: Theory. J Nondestr Eval 13(2):85–99
16. Shlivinski A, Langenberg KJ (2007) Defect imaging with elastic waves in inhomogeneous–anisotropic materials with composite geometries. Ultrasonics 46(1):89–104
17. Connolly GD, Lowe MJS, Rokhlin SI, Temple JAG (2009) Imaging of simple defects in austenitic steel welds using a simulated ultrasonic array. Rev Prog Quant Nondestr Eval 28:880–887
18. Connolly GD, Lowe MJS, Rokhlin SI, Temple JAG (2010) Synthetically focused imaging techniques in simulated austenitic steel welds using an ultrasonic phased array. Rev Prog Quant Nondestr Eval 29:871–878

19. Chatillon S, Mahaut S, Dubois P (2009) Simulation of advanced ut phased array techniques with matrix probes and dynamic settings for complex component inspections. Rev Prog Quant Nondestr Eval 28:864–871
20. Jones AT, Baxter C (1995) The rolls-royce scorpio system. Meas Sci Technol 6(1):131–133
21. Miller GF, Pursey H (1954) The field and radiation impedance of mechanical radiators on the free surface of a semi-infinite isotropic solid. Proc Roy Soc A 223(1155):521–541

Chapter 4
Crystallographic Orientation Using Ultrasonic Arrays

Abstract The anisotropic behaviour of single-crystal components significantly distorts ultrasonic images produced using an array probe, leading to unreliable inspections. Therefore, imaging algorithms must be corrected to compensate for the anisotropy. In the previous chapter, the TFM ultrasonic imaging algorithm was successfully corrected using the analytical models developed in Chap. 2. This correction requires the accurate knowledge of the velocity profile in an anisotropic material. In a single-crystal component, such as a jet-engine turbine blade, the velocity profile is controlled by its crystallographic orientation. Therefore, this orientation must be measured before the component can be inspected for defects. Conventional crystallographic orientation methods rely on X-ray diffraction. These methods are very accurate, however the size of the equipment and the requirement for the component to be etched prior to the measurement makes them unsuitable for in situ inspections. A far more expedient method is shown to be to use the same ultrasonic array that performs the defect inspection to also measure the crystallographic orientation. In this chapter, potential ultrasonic array crystallographic orientation methods are presented. The most appropriate methods are developed further and their orientation accuracy and reliability are demonstrated using simulations and experiments.

4.1 Introduction

In the previous chapter, it was shown that using the incorrect crystallographic orientation can significantly distort an image produced with the corrected TFM algorithm when inspecting a single crystal component. The crystallographic orientation of the component should be measured prior to the ultrasonic inspection so that the correct velocity profile is used. Therefore, in this chapter, potential crystallographic orientation methods are discussed and analysed.

C. Lane, *The Development of a 2D Ultrasonic Array Inspection for Single Crystal Turbine Blades*, Springer Theses, DOI: 10.1007/978-3-319-02517-9_4,

4.1.1 X-Ray Diffraction Orientation

So far, in this thesis, the orientation of single crystal test pieces has been measured using an X-ray diffraction method. Specifically, the SCORPIO system (which is an acronym for Single Crystal Orientation Rapid Processing and Interpretation Operation), developed by Rolls-Royce in the 1980s, has been used [1]. The SCORPIO system exploits the Laué/Bragg phenomenon [2]. This states that if a material is irradiated with electromagnetic radiation the particles within the material will scatter the incident rays in all directions. However, when the scattered beams are in phase, constructive interference occurs. The different arrangement of particles in the crystal lattice will cause different scattering patterns. Therefore, these scattered interference patterns can be used to identify the orientation of the crystal [3].

The SCORPIO system uses a narrow X-ray beam to irradiate the component. X-rays are used as the wavelength is similar to the spacing of atoms in a metallic crystal and this results in significant scattering. The scattering from the crystal is captured on X-ray film or using real-time detectors. The patterns are then automatically analysed on a computer using image processing to infer the crystal orientation. Just a single inspection of a clean, uncoated patch on the surface of the single crystal is required to measure the full orientation. This system is used to ensure single crystal castings have the correct crystallographic orientation prior to entering service as it is very accurate, able to measure the orientation to an accuracy of $\pm 0.8°$, and it is also rapid.

In industry, the orientation data of individual single crystal components is often recorded and stored. Therefore, the orientation of a component may be found by simply retrieving the original measurement. However, the serial numbers on the individual components are not always accessible when in-service. For such cases where the pre-service orientation measurements are not traceable, these X-ray diffraction systems are unsuitable for in situ measurements. This is due to their size and the requirement for the surface of the component to be etched to remove surface impurities and coatings. A more efficient solution is to use the same ultrasonic array that performs the bulk inspection for detecting defects to also measure the crystallographic orientation. This offers the further benefit that the orientation will always be measured relative to the array and not relative to the component's coordinates. Fundamentally this is what is required for successful ultrasonic imaging. Therefore, ultrasonic orientation methods will be considered from this point onwards.

4.1.2 Ultrasonic Orientation

It was shown in Chap. 2 that the velocity of ultrasonic waves in an elastic solid is determined by the elastic properties and the density of the material. Therefore, the determination of elastic properties using ultrasonic velocity measurements is possible and is commonly used (see Chap. 17 of Rose [4] or Sect. 6.2 of Scruby et al. [5]). In theory, the number of velocity measurements required is equal to the number of independent elastic constants. For example, the inversion of the three Eqs. (2.49–2.51) and the velocity measured along the defined crystallographic axes would yield the three elastic constants for a cubic material. However these methods require the specimen to be destructively cut in predetermined directions or using through transmission ultrasonic measurements with an oblique incident angle [6]. This latter case requires accurate knowledge of component geometry and access to two sides of the component.

Alternatively, a continuous velocity profile at different directions in the material could be measured and used to compute the elastic properties [7]. Care is required to determine whether the phase or group velocity profile is being measured. If it is the phase velocity, simple inversion of the equations describing the phase velocity profile yields the required elastic constants. However, if the group velocity profile is measured, a straightforward inversion is not always possible as closed-form expressions relating group velocity to elastic properties do not exist. Therefore, optimization procedures that minimise the error between predicted and measured group velocity profiles have been suggested [7, 8].

There are a number of different ultrasonic methods that can be used to measure the velocity profiles. For example, in 1985 Kushibiki et al. developed a line-focused acoustic microscope to measure the Rayleigh wave velocity in different directions in anisotropic crystals [9]. A similar system was developed that could also measure the surface skimming longitudinal velocity on hexagonal crystals [10]. Frénet et al. developed a conical-shaped phased array probe in 1999 to measure the velocity of surface acoustic waves, which removed the need to mechanically move the probe relative to the component [11]. Alternatively, surface wave velocity measurement systems using laser generation and detection have been developed by a number of authors [12–14]. All of these surface wave methods offer the benefit of only requiring access to a single surface.

A slightly different approach was adopted for this project. This approach requires the elastic constants to be known a priori and the velocity profile measurement is used to infer the crystallographic orientation. The different ultrasonic methods to measure the velocity profiles will be discussed in the next section. The ultrasonic systems discussed above were all developed specifically to measure velocity, whereas here a phased array probe designed for imaging will be used. Therefore, the accuracies achieved are likely to be lower than the specific velocity measurements system.

An alternative image-based method will also be developed. A similar image-based approach was investigated by Hunter et al. [15]. In this chapter the authors measured the location of elements in a flexible array using the images produced by structural features within the component on which the flexible array was placed.

The orientation accuracy of all developed methods discussed below will be found using simulated and experimental data.

4.2 Velocity Profile Measurement Methods

It may seem somewhat circuitous to measure the velocity profile to infer the crystallographic orientation which is in turn used to compute the velocity profile required for the imaging algorithm. A much simpler approach would be to measure the velocity profile and use this velocity directly in the imaging algorithm. However, it is unlikely that the velocity profile could be measured over the full range of angles that are required for imaging the component. Therefore, the measured velocity profile cannot be used directly for imaging.

The proposed ultrasonic array orientation system is shown schematically in Fig. 4.1. In the scheme, the time-of-flight of a particular propagation mode is measured from the full matrix capture (FMC) data. An initial estimate of the crystallographic orientation and the known material properties are used to compute the velocity profile relative to the ultrasonic array. The orientation convention used is described by consecutive rotations around the three <001> directions with clockwise rotations being positive, as shown in Fig. 4.2. The order in which the rotations are applied affects the final orientation, and therefore care must be taken to be consistent.

Next in the orientation scheme, a propagation model is used to predict the velocity profile needed to compute the time-of-flight data. The measured and computed time-of-flight data are then compared to find the error between them. The process is repeated with different orientations until the error is minimised. This is a three-parameter minimisation problem and was solved using an exhaustive search by looping through all possible orientations. This could be replaced in the future by more sophisticated and faster minima search methods. The orientation that yields the minimum error is then used for imaging. Similar schemes have been suggested for the measurement of elastic constants of anisotropic materials [7, 8] and for predicting the location of impacts in anisotropic plates using acoustic emission data [16]. The potential propagation modes that could be used with this scheme will now be examined.

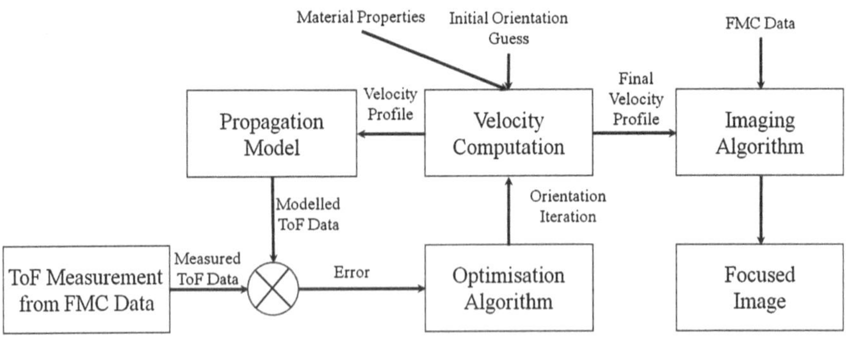

Fig. 4.1 Ultrasonic orientation method schematic

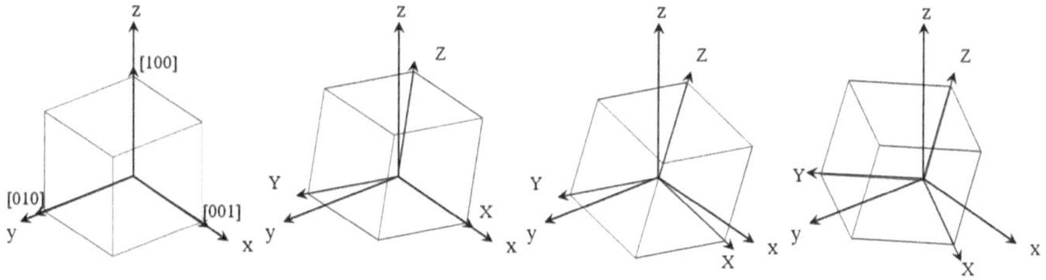

Fig. 4.2 Schematic diagram showing three consecutive rotations around the < 001 > directions

4.2.1 *Bulk Wave Methods*

4.2.1.1 Reflections off a Point-Like Reflector

The optimal method for the orientation scheme would be to measure the arrival times of bulk waves reflecting off a point-like reflector at a known location in the component under inspection. This would yield the times-of-flight of waves propagating along well-defined ray paths at a range of propagation angles. In practice, however, it is unlikely that point-like reflectors will be present in engineering components at known locations, so this method was considered not to be viable.

4.2.1.2 Reflections from a Backwall

A more common feature is a backwall surface. Therefore, reflections from this surface could potentially be used. From Fermat's principle of least travel time the ultrasonic energy will travel along ray paths that minimise the time-of-flight [17]. The time-of-flight, t, is given by:

$$t = \frac{d_1}{V(\theta_1)} + \frac{d_2}{V(\theta_2)} \tag{4.1}$$

where d is the ray path distance and V is the group velocity at propagation angle θ. The subscripts 1 and 2 indicate the incident and reflected wave respectively. For a flat isotropic sample, this means that the reflection point of an ultrasonic wave between transmit and receive elements is the geometric middle of element locations projected on to the backwall. Therefore, the incident and reflected angles must be equal. This is equivalent to having a synthetic mirror-image of the situation and the ray path transcribing a straight line from the transit element to the synthetic receive element, as shown schematically in Fig. 4.3a. The situation is slightly more complex to model in the case of anisotropic material, as beam skew due to the anisotropy will cause the ray paths to vary depending on the crystallographic orientation, as shown in Fig. 4.3b. This phenomenon has been shown in

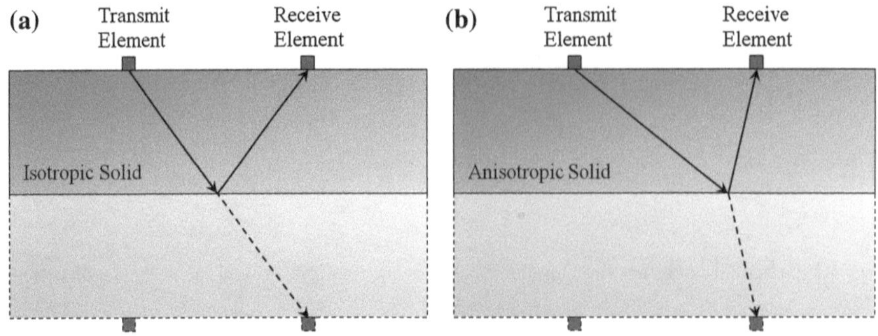

Fig. 4.3 Schematic diagram showing the ray paths between two ultrasonic transducer elements for (**a**) isotropic and (**b**) anisotropic materials

2D, but in general it is a 3D problem which must be included in the propagation model used to compute the time-of-flight data.

4.2.2 Surface Wave Method

Alternatively, the surface velocity profile could be found by measuring the time-of-flight of surface waves propagating between array elements. This method eliminates the need to have a bulk reflector, and only requires the array geometry to be known. A further benefit is that when a 2D array is used the profile is measured through a full 360°.

The computation of surface wave velocities follows a similar form to the analytical bulk wave analysis previously presented in Sect. 2.2.3. However, a boundary condition must be imposed that gives a traction-free surface. If the surface is normal to the third Cartesian axis, then the boundary condition is given by:

$$\sigma_{3i} = 0 \tag{4.2}$$

Therefore it can be shown [18] that the surface wave velocities are given by the solution to

$$det\left|C_{3jkl}u_k n_l\right| = 0 \tag{4.3}$$

One possible surface wave that could be utilised is the well-known Rayleigh wave. The Rayleigh wave exhibits both shear and longitudinal wave components, and causes particles on the surface of an elastic solid to move with elliptical motions. An important feature of these waves is that the energy is contained very close the material surface. At depths more than a few wavelengths, the displacement amplitude of particles within the material is negligible. The Rayleigh wave velocity, c_R, in isotropic materials is given by:

$$\left[2-\left(\frac{C_R}{C_s}\right)^2\right]^2=4\left[1-\left(\frac{C_R}{C_l}\right)^2\right]^{\frac{1}{2}}\left[1-\left(\frac{C_R}{C_s}\right)^2\right]^{\frac{1}{2}} \tag{4.4}$$

This equation cannot be solved explicitly to compute the Rayleigh wave velocity but may be approximated by:

$$C_R \cong C_s\frac{0.72-\left(\frac{C_s}{C_l}\right)^2}{0.75-\left(\frac{C_s}{C_l}\right)^2} \tag{4.5}$$

The ratio of longitudinal to shear wave velocities in most engineering isotropic metals is in the order of 2:1. Therefore, the Rayleigh wave velocity is approximately 90 % the speed of the bulk shear velocity in most common engineering metals.

In the anisotropic case, a trial and error procedure must be implemented that forces the determinant of Eq. (4.3) to zero. In certain directions on anisotropic surfaces, the above analysis may produce solutions that are quicker than the slowest bulk shear wave velocity. It can be shown [18, 19] that these solutions represent surface waves that leak energy into the bulk of the material. As the energy of these waves is not contained close to the surface they are not considered true Rayleigh waves but are instead named pseudo-Rayleigh waves.

Another solution to the surface wave equation is the bulk longitudinal mode that skims along the surface, often referred to as the surface skimming longitudinal wave (SSLW). The SSLW propagates at the same speed as the quasi-longitudinal bulk wave and as such, it is the first signal that appears in the time trace data, and therefore should be the easiest to time-resolve. The SSLW, therefore, also offers a potential method of computing the crystallographic orientation.

4.2.3 *Orientation Accuracy*

4.2.3.1 Experimental Set-Up

To test the accuracy of the proposed orientation methods, the following experiment was performed. A block of single crystal CMSX-4 was machined to have a wall thickness of 15 mm. The crystallographic orientation of the specimen was first measured using the SCORPIO X-ray diffraction system. The ultrasonic full matrix data was then captured using the Peak NDT Micropulse array controller with the 3 MHz 11 × 11 element 2D array placed on the surface of the test block. The captured data was then processed using the Rayleigh wave, SSLW and bulk backwall methods. As the velocities were computed between point-sources and point-receivers, it was clear that the group velocity was being measured in all cases.

The times-of-flight were measured by first windowing the individual pitch-catch time data around the time where the signal was most likely to occur. The time signals were then cross-correlated against a reference signal which was obtained from the backwall reflection in an isotropic test piece. The cross-correlation value, *CC,* is given by:

$$CC_{ij}(\Delta t) = \frac{\sum_{n=1}^{N} \left[t_{ij}(n) - \bar{t}_{ij}\right]\left[t_{ref}(n) - \bar{t}_{ref}\right]}{\left[\sum_{n=1}^{N} \left[t_{ij}(n) - \bar{t}_{ij}\right]^2 \sum_{n=1}^{N} \left[t_{ref}(n) - \bar{t}_{ref}\right]\right]^{\frac{1}{2}}} \tag{4.6}$$

where t_{ij} is the time data recorded for N samples when transmitting on element i and receiving on element j. The variable t_{ref} is the reference time signal that has been shifted by the time delay Δt. The bar indicates the mean value of the time signals. The times-of-flight were found from the time shift that caused the maximum cross-correlation value.

It was found that there was a systematic error that caused inaccuracies in the time-of-flight measurements. This error was due to a finite electronic time delay between when the array controller started timing and when the electrical pulse was actually sent from the array controller causing the array element to vibrate. The finite offset was computed for each array element by measuring the difference between the first and second arrival times of backwall reflections from an isotropic flat test block. This was repeated a number of times for each element and averaged to remove errors due to differences in the thickness of the couplant layer. It was found that average element offset for this combination of array probe and array controller was 0.58 μs. The actual times-of-flight were then computed by subtracting the offset computed for each element from the measured times-of-flight.

4.2.3.2 Experimental Results

The experimental results initially showed that the Rayleigh waves, being the slowest type of surface wave, were generally mixed in with bulk wave reflections and other reflected surface waves. In the majority of the FMC data, the Rayleigh wave signals could not be time-resolved and hence are of limited use for orientation estimation. However, for larger specimens, the Rayleigh wave method could potentially be implemented successfully.

The SSLW could be easily resolved from the data as it was the first wave signal to appear in the time trace. The bulk wave reflected from the backwall surface could also be resolved as it was the only bulk reflector in the test block, and hence could not be confused with other signals. Figure 4.4 shows the measured times-of-flight of both wave modes from the collected full matrix data. The SSLW data can be seen to be very noisy in the diagonal of the matrix which corresponds to measurements taken from the elements closest to one another. This is because the SSLW in this case has not had time to separate from the ring down in the collected data, which is caused by local mechanical resonances and electrical interference.

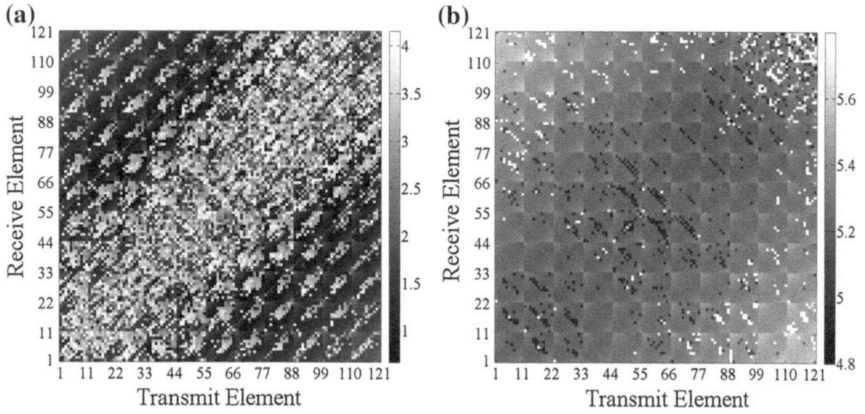

Fig. 4.4 The measured times-of-flight in microseconds for (**a**) the SSLW and (**b**) the bulk wave reflected from the backwall surface

The bulk data is a lot less noisy as the waves have travelled further and thus occur in the time trace at points where the level of noise is much lower. Certain measured time-of-flight data was not included in the analysis if it had a low cross-correlation value or was outside the expected time-of-flight range.

Table 4.1 shows the orientation measurements obtained using the SCORPIO X-ray diffraction method, the SSLW method and the backwall method. The orientation is described by three consecutive rotations around the $< 001 >$ crystallographic directions, as shown in Fig. 4.2. In fact, in both the SSLW and backwall cases, there were two minima locations, as shown later in Fig. 4.6a, c. The two minima locations correspond to symmetrically opposite orientations that yield velocity profiles that are symmetric and hence give the same time-of-flight data. The results shown in Table 4.1 are the orientations that most closely match the SCORPIO results.

Table 4.1 shows that the SSLW results are most closely correlated to the SCORPIO results despite the high level of noise in the measured data. The closest correlation is for the rotation around the Z axis measurement. In this example, the Z rotation relates to the orientation of the crystal in the plane of the ultrasonic array.

Table 4.1 The orientation measurements found using the SCORPIO X-ray diffraction method, the SSLW method, and the backwall method

Orientation	Rotation around X axis (°)	Rotation around Y axis (°)	Rotation around Z axis (°)
SCORPIO	−5.2	−17.4	−1.4
SSLW method	2	−16	−2
Backwall method	−9	−21	1

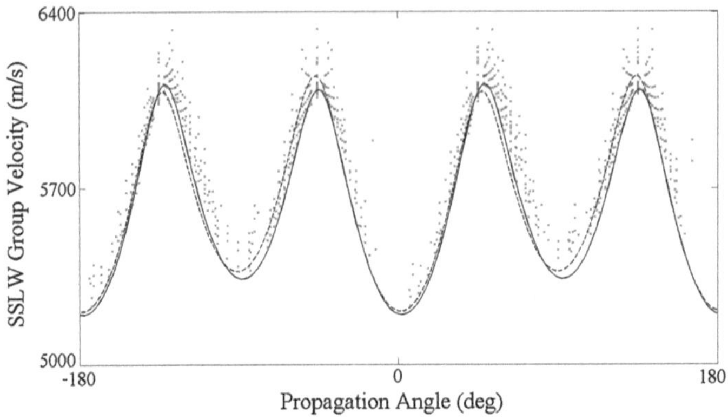

Fig. 4.5 The measured SSLW group velocity profile (crosses) and the modelled velocity profile from the ultrasonic SSLW orientation method (*solid line*) and SCORPIO X-ray diffraction method (*dashed line*)

Figure 4.5 shows the measured surface velocity profile of the SSLW. The SSLW orientation method maximises the correlation by minimising the error between the measured velocity profile and the modelled profile, by adjusting the latter (shown as a solid black line). The rotation around the Z axis has the effect of shifting the entire modelled profile along the propagation angle axis, while the other two rotations cause the relative heights of the peaks and troughs of the modelled profile to vary. The error is very sensitive to variations in the Z rotation, as shown in Fig. 4.6b, and this explains why the best results are obtained for this rotation. The error is much less sensitive to the other two rotations, as shown in Fig. 4.6a, where the global error is fairly constant over a large range of rotation angles. Therefore although the SSLW method seems to be accurate for the out-of-plane rotations, particularly in the measurement of rotation around the Y axis, these results can be considered less reliable than the measurement of rotation around the Z axis.

Figure 4.6c shows the global error between the measured and modelled data for the backwall method for the rotations around the X and Y axes whilst the Z rotation is kept constant. A distinct circular pattern of minima can be observed. This circular pattern represents an orientation that relates to constant velocity in the direction normal to the plane of the array. The reason for this minima pattern is most likely because the bulk method only considers waves that have travelled under an acute range of propagation angles between the array and the backwall surface. Therefore, small changes in these two orientations will only significantly change the computed error if they affect the velocity profile in this small range of propagation angles. The circular minima pattern does not intersect the true orientation due to the experimental variations. For example, the exact location and shape of the backwall surface may have varied from the model leading to inaccuracies. Further inaccuracies could have been introduced when modelling the beam skew caused by the anisotropy.

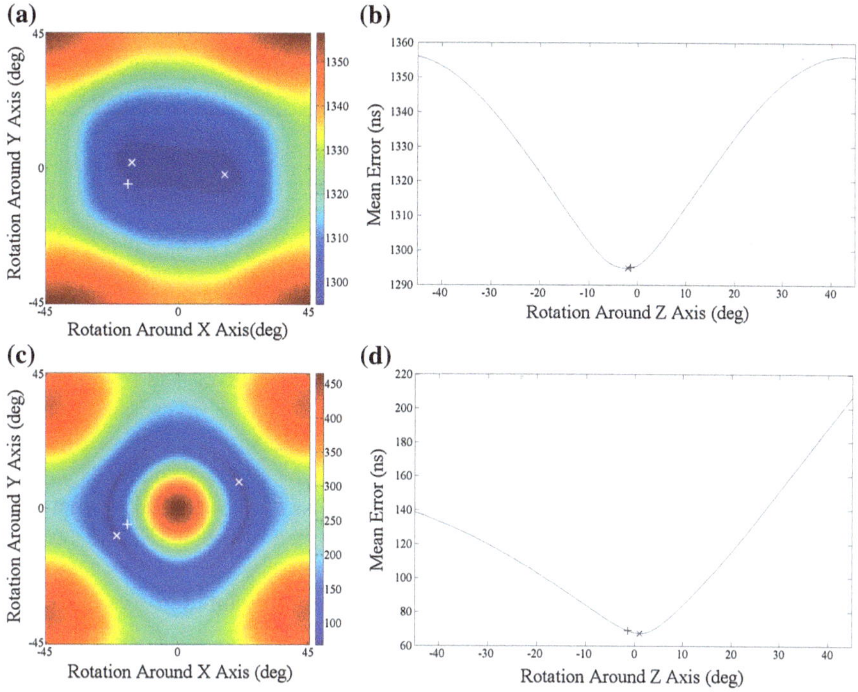

Fig. 4.6 The average error between the measured time-of-flight data and the computed time-of-flight data for rotations around **a** the X and Y axes and **b** the Z axis for the SSLW method and **c** the X and Y axes and **d** the Z axis for the backwall method. The plots **b** and **d** are shown using the rotations X and Y that were found to give the minimum error in plots **a** and **c** respectively. The + shows the orientation found using the SCORPIO system and the × indicates the minimum error in each plot

The first experiment showed that the SSLW method offered the best accuracy in determining the orientation of a single crystal but only in the plane of the array. Therefore, a second experiment was performed to investigate whether measurements on multiple surfaces of the components could yield the full orientation. The experiment was performed as follows: FMC data was collected from the two surfaces orthogonal to the original inspection surface. The SSLW method was then used to process the data to find the orientation in the plane of the individual surfaces. In this way the full three rotations could be found from the three orthogonal surfaces. Table 4.2 shows the result of this experiment. It can be seen that this method is able to measure the complete orientation to an accuracy of $\pm 2°$. The backwall method was not used as this would require a rectangular test specimen. This would therefore have very little applicability to real engineering structures. Conversely, access to three orthogonal surfaces, required for the SSLW method, is feasible.

Table 4.2 The orientation measurements found using the SCORPIO X-ray diffraction method and the SSLW method using three orthogonal surfaces

Orientation	Rotation around X axis (°)	Rotation around Y axis (°)	Rotation around Z axis (°)
SCORPIO	−5.2	−17.4	−1.4
SSLW method	−5	−19	−2

4.2.3.3 Simulated Accuracy

The experimental results showed that none of the velocity profile measurement methods could reliably find an accurate orientation of a single crystal from a single inspection location. Therefore, simulations were performed to investigate whether the full orientation could potentially be found from velocity measurements with minimal errors.[1]

Time-of-flight data was simulated for both the backwall and SSLW methods using the orientations of the real test piece as determined by the SCORPIO system. The simulated data was then processed using the orientation scheme assuming there is access to only a single surface. The results of the simulated experiments are shown in Fig. 4.7. It can be seen that the shape of the error fields closely matches the experimental results. However, the minimum error for the simulated results approaches zero, whereas the experimental errors are much larger due to experimental inaccuracies. The simulated results do not actually reach zero error because the orientation measurements were only processed in one degree increments, whereas the time-of-flight data was simulated using orientations with fractions of degrees. As with the experimental results, there are two minima locations for both simulated results corresponding to symmetrically opposite orientations that yield the same time-of-flight data. It can be seen that even with zero error in the velocity measurements there is still ambiguity in the rotations around the X and Y axes. Therefore, it is unlikely that the single-surface measurement methods could be practically used to measure the full orientation of a single crystal as there will always be a small amount of velocity measurement error or ambiguity in real experiments.

Figure 4.8 shows the average error for rotations around the Y and Z axes for the SSLW and backwall methods. The minimum error at each Y rotation is indicated with a white dot. For the SSLW method, the minimum error for the rotation around the Z axis does not deviate from the true orientation by more than 1° as the rotation

[1] When measuring the velocity in this way the error will never be zero even if the time-of-flight data is completely accurate. This is because waves propagate from a point-like source, such as an ultrasonic array element, as spherical waves. The spherical wave causes a stiffening of the material causing an increase in the velocity. Once the waves have propagated a couple of wavelengths, the spherical waves become more like plane waves and therefore this effect diminishes with distance from the point-like source.

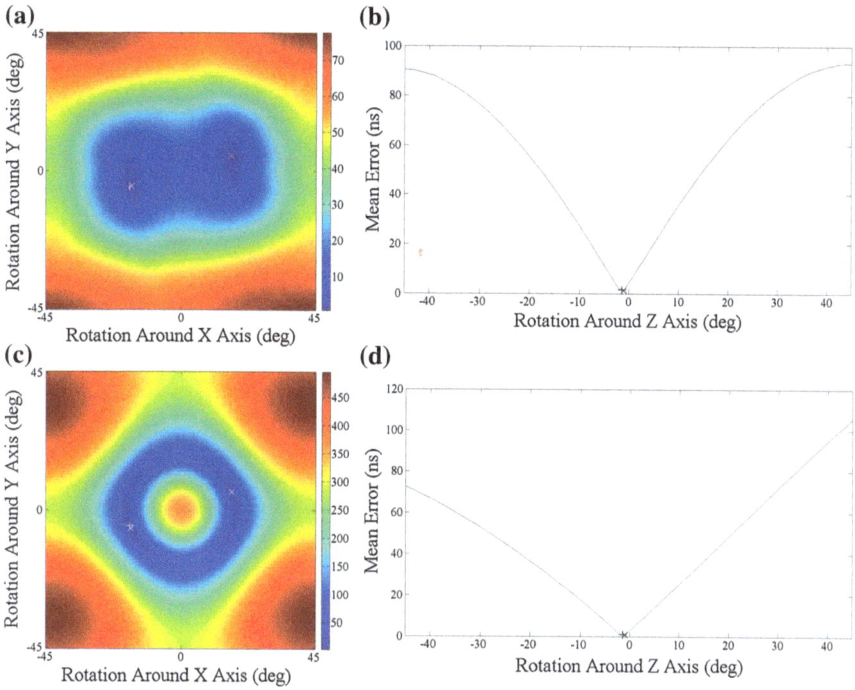

Fig. 4.7 The average error between the simulated time-of-flight data and the time-of-flight data computed for rotations around **a** the X and Y axes and **b** the Z axis for the SSLW method and **c** the X and Y axes and **d** the Z axis for the backwall method. The plots **b** and **d** are shown using the rotations X and Y that were found to give the minimum error in plots **a** and **c** respectively. The + shows the orientation found using the SCORPIO system and the × indicates the minimum error in each plot

around Y is varied, as shown in Fig. 4.8a. The determination of the rotation around the Z axis is therefore almost decoupled from the determination of the other rotations, and can be considered in isolation. This could significantly speed up the optimization process to find the Z rotation. However, by examining Fig. 4.8b, it is clear that this is not the case for the backwall method.

4.3 Image-Based Method

It was demonstrated in the previous section that none of the velocity profile based methods could successfully determine the X and Y orientations from a single inspection location. However, it was also shown in Sect. 3.5 that these two orientations have a significant effect on the TFM image, as these caused the largest change in the velocity profile used for imaging. Therefore, an alternative image-based method was investigated. This method works by comparing images

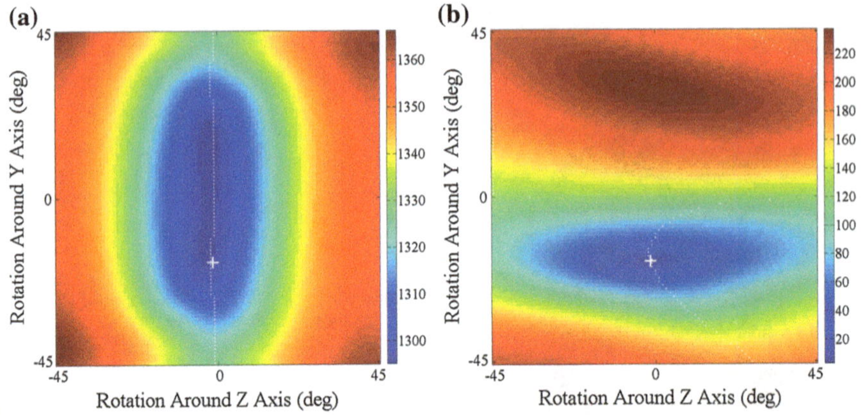

Fig. 4.8 The average error between the simulated time-of-flight data and the times-of-flight computed for rotations around the Y and Z axes for (**a**) the SSLW method and (**b**) the backwall method. The + shows the true orientation and the dots indicates the minimum error at each Y axis orientation

reconstructed using various trial crystallographic orientations to a reference image. The reference image could be from either an isotropic test piece with the same geometry, or from a single crystal component whose orientation is known from X-ray diffraction measurements. The correct orientation can then be found when the reconstructed image has the highest correlation with the reference image. This does assume that the geometry of the components under inspection does not vary significantly. This is reasonable in the case of turbine blades which are manufactured to very exacting tolerances.

4.3.1 Orientation Accuracy

A simulation of the image-based data could be achieved using a 3D FE model. However, this was not performed, as an FE model with a fine enough mesh to yield a reasonable accuracy would require more computer memory than was available for this project. Therefore, to test the accuracy of this method, the following experiment was performed. Specimens with identical geometries were machined from isotropic nickel alloy and from single crystal CMSX-4. The geometry of the specimens was derived from the cross-section of a typical turbine blade root. Figure 4.9a shows a diagram of an array probe on the platform of a turbine blade performing an inspection of the blade root. The cross-sectional view of this arrangement is shown in Fig. 4.9b. In this image the region in bold shows the geometry that was used for the isotropic and single crystal test pieces. The full matrix of data from both specimens was captured using a 12 × 6 element 10 MHz array with a 0.6 mm pitch, manufactured by Sonaxis (Besançon, France).

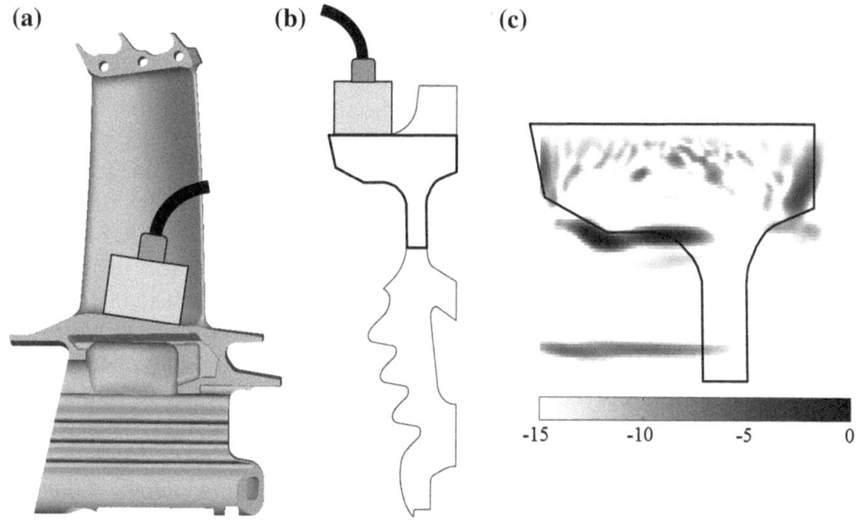

Fig. 4.9 **a** A schematic diagram showing an array probe placed on a turbine blade to perform a blade root inspection, **b** the cross-sectional view of the probe on the blade, with the line in bold showing the geometry used for the image-based orientation specimens, and **c** the isotropic TFM y-z plane image used for the reference image shown using the dB scale. Image **a** courtesy of Rolls-Royce plc

The 3 MHz Imasonic array was too large to be used for this test piece. To begin, the TFM image of the isotropic test piece was computed and this image was used as the reference image. To reduce computation time, instead of computing a full volumetric image, a slice from each of the x-z and y-z planes were used. An example of the reference TFM slice through the y-z plane is shown in Fig. 4.9c. Next, images of the CMSX-4 specimen were computed at different orientations. At each orientation the reconstructed image was cross-correlated against the reference image. It was assumed that the orientation controlled by the rotation around the Z axis had first been measured using the SSLW method. In addition, Sect. 3.5 showed that this rotation had very little effect on the TFM image. Therefore, only the remaining two orientations were varied.

Table 4.3 shows the orientation of the single crystal specimen determined using the SCORPIO system and the image-based method. It can be seen that the image-based method was within ±1.2° of the SCORPIO orientation.

Table 4.3 The orientation measurements found using the SCORPIO X-ray diffraction method and the image-based method

Orientation	Rotation around X axis (°)	Rotation around Y axis (°)
SCORPIO	−0.4	0.2
Image-based	0	−1

Figure 4.10 shows a plot of the cross-correlation coefficient between the isotropic reference images and the TFM images from the single crystal computed as the X and Y rotations were varied. The cross-correlation was performed on images using the dB scale. The images were normalised so that the maximum platform signal in the reference image was 15 dB. Pixel values less than 0 dB were set to equal 0. The purpose of this threshold was to remove the effect of image noise on the cross-correlation. Marked on the plot is a white cross which shows the SCORPIO orientation. It can be seen that the cross-correlation value increases rapidly in the area of the SCORPIO orientation. Therefore, the image-based method was shown to be accurate at measuring the two out-of-plane orientations. However, this result was only based on one laboratory experiment, in which the position of the array relative to the test specimen was tightly controlled using a positioning jig. An in situ measurement may not be as accurate due to variations in the probe positioning and the geometry of the individual blades. This will be investigated in Sect. 5.4.5.

In addition, it should be noted that a relatively complex geometry should be used for the image-based orientation method. To demonstrate, if a component only has a flat backwall to use for the correlation, then the results will not be any better than the velocity profile based method. This is because the two methods are fundamentally the same, in that they are both attempting to find the velocity profile that best matches the recorded time-of-flight data. In the velocity profile based approach, a simple and fast ray-tracing model, using a priori information (i.e. the location of the backwall), is implemented. However, the TFM imaging algorithm effectively uses ray-tracing to find whether there is a point-like scatter at every pixel location within the image, and as such is much slower. The advantage of using the image-based method is that complex geometries cannot be accurately modelled using ray-tracing methods but they can be imaged.

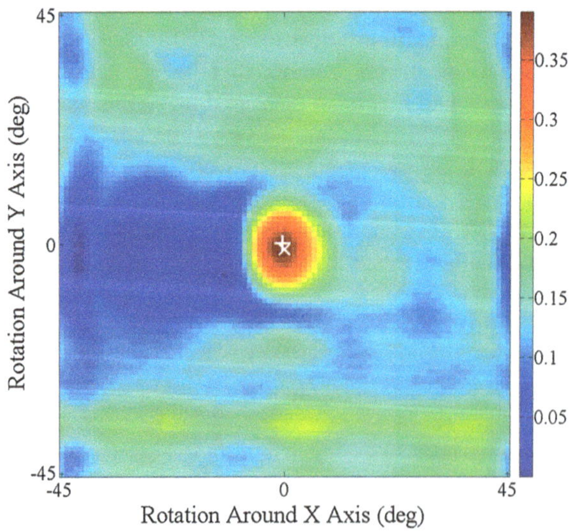

Fig. 4.10 The cross-correlation coefficient between the isotropic reference images and the TFM images from the single crystal computed, as rotations X and Y were varied. The cross indicates the location of the maximum cross-correlation coefficient and the + indicates the orientation found from SCORPIO measurements

4.4 Summary

A number of potential crystallographic orientation measurement methods using an ultrasonic array have been investigated in this chapter. From experimental results, none of the velocity profile measurement methods were found to be able to reliably predict the full orientation of a single crystal. However, the SSLW wave was able to accurately measure the orientation of the crystal in the plane of the array. Therefore, by measuring the SSLW velocity of three mutually orthogonal surfaces the full orientation could be found. Simulated results show that even with a minimal velocity measurement error none of these proposed methods could predict the full orientation from a single inspection location. However, an image-based method was shown to be able to accurately determine the two out-of-plane orientations.

All of the results presented in this chapter are only applicable to the specimen and array geometries used for the particular experiments mentioned. Different accuracies could be achieved with different experimental set-ups, and orientation accuracies should be computed for specific inspections. Therefore, in the following chapter, in which an ultrasonic inspection system is developed for a specific turbine blade inspection, the analysis described above is repeated for the new scenario.

References

1. Jones AT, Baxter C (1995) The rolls-royce scorpio system. Measur Sci Technol 6(1):131–133
2. Woolfson MM (1997) An introduction to X-Ray crystallography, 2nd edn. Cambridge University Press, Cambridge UK
3. Stout GH, Jensen LH (1989) X-ray structure determination, 2nd edn. Wiley, New York
4. Rose JL (1999) Ultrasonic waves in solid media. Cambridge University Press, Cambridge
5. Scruby CB, Drain LE (1990) Laser ultrasonics techniques and applications. Adam Hilger, Bristol
6. Markham MF (1970) Measurement of the elastic constants of fibre composites by ultrasonics. Composites 1:145–149
7. Degtyar AD, Rokhlin SI (1997) Comparison of elastic constant determination in anisotropic materials from ultrasonic group and phase velocity data. J Acoust Soc Am 102(6):3458–3466
8. Every AG, Sachse W (1990) Determination of the elastic constants of anisotropic solids from acoustic-wave group-velocity measurements. Phys Rev B 42(12):8196–8205
9. Kushibiki J-I, Chubachi N (1985) Material characterization by line-focus-beam acoustic microscope. IEEE Trans Sonics Ultrason 32(2):189–212
10. Kielczynski PJ, Bussière JF (1991) Characterization of texture in hexagonal materials using a line focus acoustic microscope. In: Ultrasonic symposium proceedings IEEE, pp 1009–1031
11. Frénet D, Calmon P, Paradis L, Roy O, Ouaftouh M (2000). Modeling of surface acoustic waves reflected on fluid-loaded solids. Application to measurements on anisotropic materials with a phased-array broadband transducer. In: Rev prog Nondestr Eval, vol 19, p 1017–1024

12. Robbins WP, Rudd EP (1987) Measurement of surface acoustic wave velocity anisotropy with a scanning laser acoustic microscope. In: Ultrasonics symposium proceedings IEEE, p 785–789
13. Reverdy F, Audoin B (2001) Elastic constants determination of anisotropic materials from phase velocities of acoustic waves generated and detected by lasers. J Acoust Soc Am 109(5):1965–1972
14. Hong Y, Sharples SD, Clark M, Somekh MG (2004) Rapid and accurate analysis of surface and pseudo-surface waves using adaptive laser ultrasound techniques. Ultrasonics 42(1–9):515–518
15. Hunter AJ, Drinkwater BW, Wilcox PD (2010) Autofocusing ultrasonic imagery for non-destructive testing and evaluation of specimens with complicated geometries. NDT E Int 29(2):78–85
16. Kundu T, Das S, Kumar VJ (2007) Point of impact prediction in isotropic and anisotropic plates from the acoustic emission data. J Acoust Soc Am 122(4):2057–2066
17. Connolly GD, Lowe MJS, Rokhlin SI, Temple JAG (2009) The application of Fermat's principle to the improved inspection of austenitic steel welds. Proc Roy Soc A 465:3401–3423
18. Farnell GW (1978) Types and properties of surface waves. In: Oliner AA (ed) Topics in applied physics. Springer, Berlin, pp 13–59
19. Briggs A (1992) Acoustic microscopy, in monographs on the physics and chemistry of materials. Oxford University Press, Oxford

Chapter 5
The Development of an in Situ Ultrasonic Array Inspection System

Abstract This chapter details the development of a complete ultrasonic array imaging system for the inspection of a specific single-crystal turbine blade. The system is developed to detect cracks that have initiated on the internal surface of the turbine blade root. The root section of the turbine blade experiences the highest operational stress and is therefore most prone to in-service cracking. The system is also designed to perform in situ inspections to avoid the significant time and cost implications of removing the turbine blades from the engine prior to inspection. As such, the design of the system is constrained by the limited access to the turbine blades within the engine. The system utilises the corrected imaging algorithm described in Chap. 3 and the ultrasonic crystallographic orientation method developed in Chap. 4. The developed system is shown to be able to successfully detect artificial defects in turbine blades. The detection performance of the ultrasonic array system as the size and orientation of defects varies is demonstrated on test specimens. In addition, the sensitivity of the inspection system due to variations in beam amplitude and misalignment in the crystallographic orientation are also evaluated experimentally.

5.1 Introduction

In Chap. 2, the theory of wave propagation in anisotropic media was presented. Analytical models were developed to compute the velocity and beam amplitudes of ultrasonic waves from a point-force on the surface of an anisotropic solid. It was then shown, in Chap. 3, how the TFM imaging algorithm could be corrected for imaging anisotropic materials provided the orientation of the material is known. In Chap. 4, methods to measure the crystallographic orientation of anisotropic single crystal components were described and evaluated. In all of these three chapters the analysis was performed using single crystal nickel-based superalloy CMSX-4. However, the experimental set-ups were fairly generic and idealised.

C. Lane, *The Development of a 2D Ultrasonic Array Inspection for Single Crystal Turbine Blades*, Springer Theses, DOI: 10.1007/978-3-319-02517-9_5,

From this point on in the thesis, the development of a specific ultrasonic array inspection will be considered. The work that has been performed in the previous chapters will be utilised for this development. In this chapter, the inspection situation will be described and an inspection system will be designed to meet the specifications. In Chap. 6, the developed inspection system will be evaluated for its defect detection and sizing capabilities.

5.1.1 Inspection Specification and Strategy

The specific inspection concerns an experimental high-pressure stage-1 turbine (HPT1) blade from a two-spool demonstrator turbofan engine. The turbine blade is cast from CMSX-4. Figure 5.1 shows CAD drawings of this experimental turbine blade. It was found that after approximately 2,000 cycles in a test-bed engine, cracks were beginning to initiate on the internal cooling passage in the root-section under the trombone, as shown in Fig. 5.1a. The crack would then propagate through the shank wall to the external blade surface, as shown in Fig. 5.1b and c. At this stage the cracks were found to be approximately 3–5 mm in length. The crack could then potentially grow from fatigue, until catastrophic rupture and failure of the turbine blade occurs.

The inspection system was designed to detect and characterise the cracks whilst they were in the primary failure stage (from initiation on the internal surface to propagation to the external surface). The inspection was required to be performed in situ, to avoid the cost associated with removing the turbine blades from the engine (see the discussion in Sect. 1.1). As the defects occur in a location with no direct access, it is clear that an ultrasonic inspection is the only feasible method (see Sect. 1.2.2).

The ultrasonic inspection strategy is shown schematically in Fig. 5.2. A 2D array will be placed on the turbine blade platform. The array will then send an ultrasonic wave towards the root shank wall where cracks are likely to initiate.

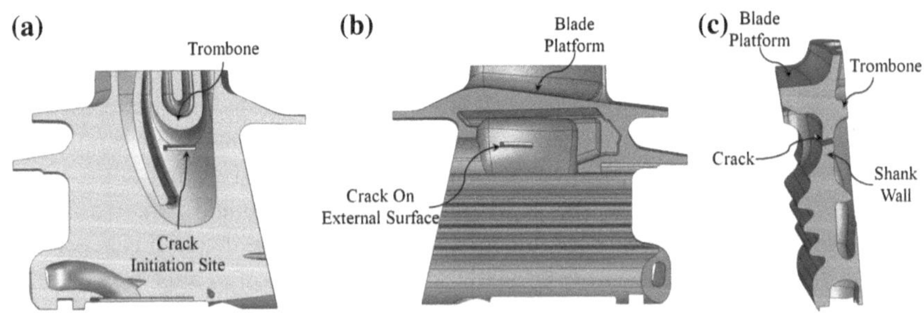

Fig. 5.1 CAD drawings of a sectioned HPT1 blade showing **a** the crack initiation site on the internal surface of the blade root under the trombone, **b** the external surface of the blade, and **c** a cross-section of the blade root shank wall. Images courtesy of Rolls-Royce plc

Any defects will scatter the incident wave and the reflected waves can be detected using the array. The detected reflected waves may be either from a direct reflection, or a skip off the defect and shank wall, as shown in Fig. 5.2b. A 2D array was chosen, as steering was required in all three dimensions, as the exact defect location and orientation may change from blade to blade.

A wedge design was considered to physically steer the ultrasonic energy into the shank region. However, as the discussion in the next section will show, the total probe height had to be less than 8 mm. This would mean that the wedge would have to be small and this would cause a high amount of reverberation in the wedge. This, in turn, would lead to an unacceptable level of interference with the signals from the turbine blade geometry and potential defects. Therefore, a direct contact inspection was chosen.

The design of the probe deployment system and that of the ultrasonic array probe will now be described in turn. A second, more conventional probe was also developed using just a single ultrasonic element in a pulse-echo configuration. The development of this single element probe is discussed in Appendix D. This probe was developed so that the phased array inspection could be compared directly with a more commonly used single element inspection.

5.2 The Development of the Probe Deployment and Manipulation System

The probe deployment and manipulation system to position the array probe on the HPT1 blade platform was developed by the on-wing inspection team from Rolls-Royce Deutschland. Therefore, only a brief description of the system will be given here.

To access the turbine blades inside the engine, the ultrasonic probe must pass through the outer-casing of the engine, into a hole left by removing a fuel sprayer

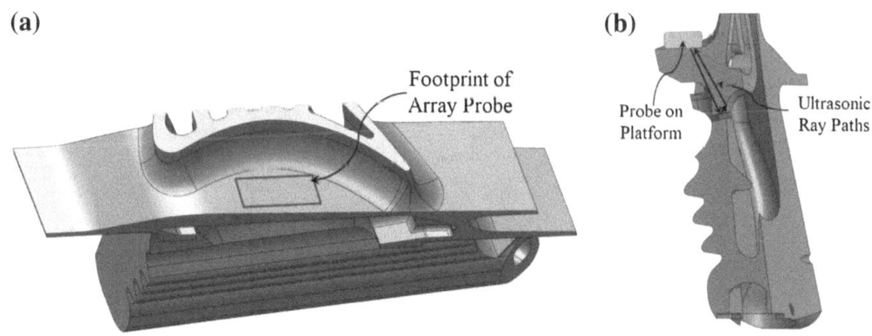

Fig. 5.2 Schematic diagrams of the ultrasonic array inspection strategy showing **a** the location of the ultrasonic array probe, and **b** potential ray paths. Images courtesy of Rolls-Royce plc

and then into the combustion chamber. From the combustion chamber the probe passes through the nozzle guide vanes (NGVs) to arrive at the HPT1 blade. This is shown schematically in Fig. 5.3. To fit between the NGVs, the probe height could not be more than 8 mm. In addition, to fit between the individual turbine blades the probe could not be more than 20 mm long and 6 mm wide.

To achieve the required access, the ultrasonic probe was attached to the end of a handheld mechanical manipulator made by Everest VIT (Henchen, Germany). Views of the probe inside the engine are provided by a flexible videoscope made by Olympus (Tokyo, Japan). Attached to the videoscope is a tube for applying ultrasonic couplant to the turbine blades.

During development it was found that the probe manipulator could not provide sufficient downwards pressure of the array probe on to the blade platform. Therefore, an inflatable balloon was designed to sit above the probe manipulator between the NGVs. The balloon stopped the probe manipulator cable from *bowing* through the NGVs and provided a resistive force so that the appropriate amount of pressure on the array probe could be achieved.

5.3 The Design of the Ultrasonic Array Probe

5.3.1 General Ultrasonic Array Design Considerations

The performance of an array, in terms of resolution, focusing and inspection coverage, increases with aperture size [1]. This requires the array to be composed

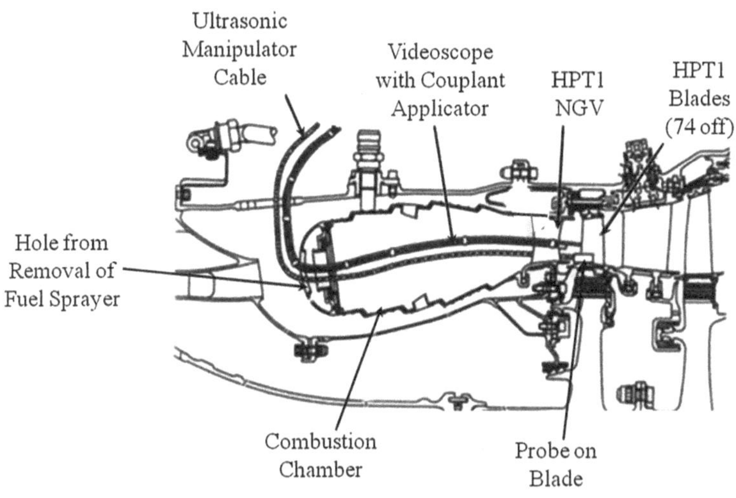

Fig. 5.3 Schematic diagram showing access route to the HPT1 blades. Image courtesy of Rolls-Royce plc

of many elements, because if the element pitch is too large, grating lobes can occur. Grating lobes send ultrasonic energy away from the intended focusing direction of the main beam. These can cause image artefacts that can potentially interfere with defect signals. For linear and matrix arrays, the element pitch should be less than half a wavelength to avoid grating lobes and the possible complications that they can cause. This, however, assumes that the array elements are perfect point-like sources. In reality elements tend to be more directional due to their finite size, and therefore grating lobes will be minimised if the element pitch is close to the element size, and the steering angle is not too great.

For linear arrays there is only one possible element layout for regularly spaced elements. However, for 2D arrays an infinite number of regularly spaced element layouts are possible. The most common is a regular grid pattern. Although hexagonal patterns have been shown to offer the greatest aperture size over all regular layouts for a given number of elements, and maximum grating lobe suppression. However, non-uniform patterns have been shown to offer twice the aperture size of the equivalent regular pattern. This is because the non-uniform arrays effectively spread noise from grating lobes throughout the entire image [2].

The near-field extent, N, of a 2D array is given by:

$$N < \frac{L^2}{\lambda} \tag{5.1}$$

where L is the array aperture size and λ is the ultrasonic wavelength in the material under inspection. If imaging in the far-field, it has been shown that sparse arrays (for example using just the perimeter elements of a grid array) can give the same imaging performance as a fully populated array [2]. This can decrease the complexity of the array design, and decrease computational time as there is less data to process. Therefore, when a 2D array probe is designed, the near-field extent should be considered.

5.3.2 Initial Ultrasonic Array Experiments

For the in situ inspection, the aperture of the array is limited by the area of the blade platform on which elements can be placed. This is approximately 11 mm by 5 mm. Furthermore, current manufacturing capability limits the element size to be no smaller than 0.5 mm. At this size, only square elements arranged on a grid pattern could be successfully manufactured and incorporated into a working array probe, at the time of writing.

The relationship between ultrasonic frequency, f, velocity, c, and wavelength, λ, is given by:

$$f = \frac{c}{\lambda} \tag{5.2}$$

Therefore, as the element pitch should be less than half the ultrasonic wavelength to avoid grating lobes, and assuming the velocity in the blade is 5 km/s, ideally a centre frequency of less than 5 MHz should be used. However, initial experiments with probe manufacturer Sonaxis, showed that at 5 MHz the electrical impedance of the array elements was too high, resulting in poor sensitivity and long ring downs. Therefore, the decision was made to design the probe to operate at 10 MHz, which also improves the imaging resolution, whilst being aware that grating lobes could occur at high steering angles.

From Eq. (5.1), using a wavelength of 0.5 mm and an array aperture of 5 mm the near-field extends to a depth of 50 mm. As the inspection will image to a maximum depth of 10 mm, it is clear the imaging will always occur in the near-field of the array. Therefore, a fully populated array is required.

An initial probe, too large to fit in the engine, was commissioned from Sonaxis to assess their manufacturing capability and to aid the design of the engine probe. This probe had 72 square elements in a 6×12 regular grid arrangement. Each element was 0.5 mm in length with a 0.6 mm pitch between adjacent elements. The centre frequency of the elements was 10 MHz.

To investigate the inspection capability of the probe the following experiment was performed. The array probe was placed on test pieces that were representative of the turbine blade cross-section, shown in Fig. 5.4. These are the same design of test piece used for the image-based orientation experiments detailed in Sect. 4.3. The test pieces were made of isotropic nickel to remove the complications of inspecting anisotropic material. The different pieces contained through-going slots, 3 mm in length, at different depths in the shank wall. These slots are at locations which correspond to where cracks had been found in the experimental turbine blades. The centre of the probe was positioned directly over the centre of the slot and then progressively moved back. At each location the peak amplitude of the TFM response from the slot was computed.

Figure 5.5 shows TFM images of the test piece with no slot and with a slot at a depth of $z = 4.5$ mm at a probe displacement of $\Delta y = 2.5$ mm. By comparing the two sets of images, the slot indication can be easily identified. The mode-converted platform signal denoted in Fig. 5.5a comes from the longitudinal wave

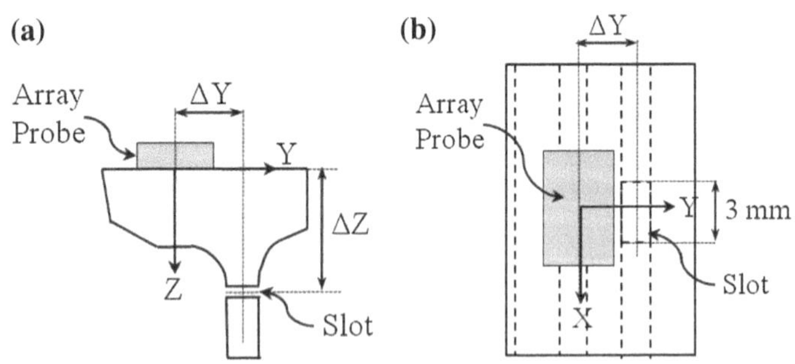

Fig. 5.4 Diagram showing the inspection of the cross-section with a through-going slot, 3 mm long in the x direction. The total depth of the test piece in the z direction was 10 mm

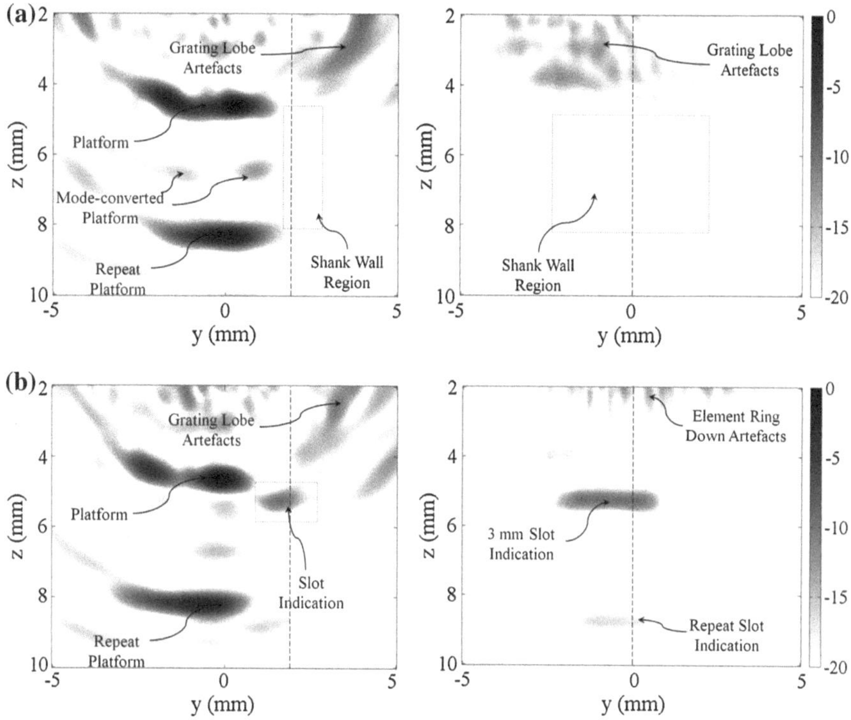

Fig. 5.5 TFM images of the cross-section test pieces in the y–z plane at x = 0 mm (*left*) and x–z plane for y = 2 mm (*right*) for a probe displacement of Δy = 2.5 mm. Images are shown for the specimen with **a** no slot and **b** a 3 mm long slot at a depth of z = 4.5 mm. Each image is displayed on a dB scale with 0 dB set to the maximum signal of the platform. The vertical dashed-lines indicate the position in which the image slices have been computed

mode-converting to a shear wave when reflecting off the platform surface. In addition, artefacts from grating lobes of the platform can be seen in the images. However, the artefacts do not occur in a region where defects could be present and therefore do not interfere with the defect detection.

Figure 5.6 shows the response from the slots located at depths of 4.5, 6 and 8 mm, as the displacement between the centre of the probe and the centre of the slot, Δy, increased. The results have been normalised to the maximum signal from the 4.5 mm deep slot at $\Delta y = 0$ mm. It can be seen that the signal response for all notches falls rapidly when Δy increases beyond 1 mm. This is because at this distance, the ultrasonic energy is no longer being scattered from the notch directly back to the array, but skips from the notch to the shank wall and then back to the array. The signal response when no slot was present, hence the noise level, is also displayed on the graph. At $\Delta y > 3.5$ mm the slot response is indistinguishable from the noise. For the real turbine blade, the blade aerofoil does not allow the displacement to be less than 2.8 mm which is indicated on Fig. 5.6 by a vertical

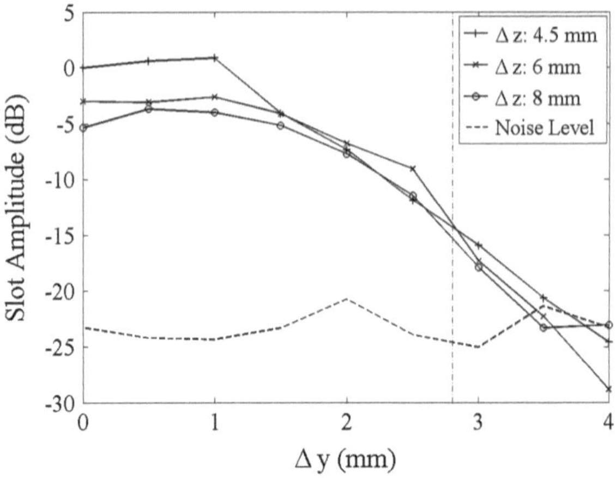

Fig. 5.6 The signal response from the slots as Δy (the distance between the centre of the array and centre of the slot) increased. 0 dB was set to equal the 4 mm deep slot amplitude at $\Delta y = 0$ mm

dashed line. At this displacement, the signal response from the slots is about 10 dB greater than the recorded noise. This shows that defect detection with a similar design of probe should be possible.

5.3.3 Design of the in Situ Engine Probe

The next probe that was manufactured was designed so that it could fit inside the engine. CAD drawings of this probe are shown in Fig. 5.7. The shape of the probe housing has been designed so it replicates the geometry of the turbine blade aerofoil, which allows the array elements to be as close to the aerofoil as possible. The probe is made up of 63 imaging elements. The element layout has been changed from a simple rectangular grid, to allow the elements to follow the radius of the aerofoil and to push the front elements as far forward as possible.

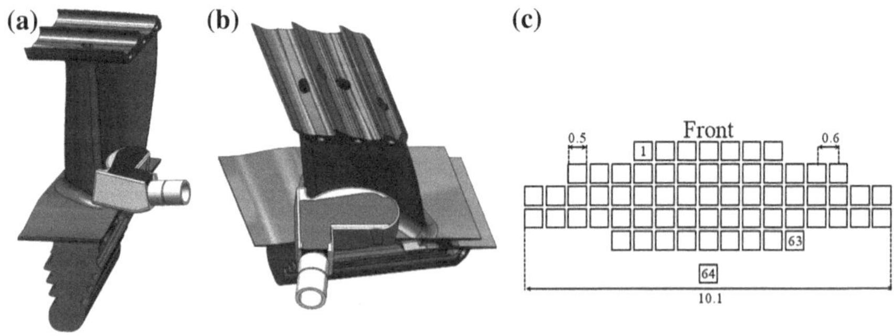

Fig. 5.7 Drawings showing **a** and **b** the engine probe on the turbine blade, and **c** the element layout. All dimensions are shown in units of mm. Images courtesy of Rolls-Royce plc

5.3.4 Crystallographic Orientation Considerations

It was found that the surface skimming longitudinal waves (SSLW), which could potentially be used for crystallographic orientation, could not be resolved from the time traces as the elements were too close and the SSLW signal was not sufficiently separated from the element ring down. Therefore, a separate element was designed to transmit the SSLW to the array elements. The element location is indicated as element 64 in Fig. 5.7. This element underwent a number of design changes. For example, the frequency was changed from 10 to 5 MHz as the array system exhibited high electrical ring down at 10 MHz. In addition, a cork barrier and electrical shield had to be incorporated around the SSLW element to minimise the mechanical and electrical cross-talk between this element and the array. This design was successful at transmitting and receiving the SSLWs.

To produce an accurate surface velocity profile, the SSLW element had to be calibrated to find its exact position relative to the array elements, and the time-delay offset caused by the polymer delay-line and the finite offset in the array controller. This was performed by collecting the SSLW data from a sample of isotropic aluminium, and then performing a multi-parameter optimization routine to minimise the difference between the measured and actual surface velocity.

The surface velocity profiles measured by the calibrated probe for isotropic aluminium and anisotropic single crystal CMSX-4 are shown in Fig. 5.8. Results that had a low cross-correlation value or were outside the expected time-of-flight range were excluded, therefore, not all 63 velocity measurements are included in these graphs. However, some outlying velocity measurements can be seen, particularly in Fig. 5.8b. This is where the cross-correlation process has selected the incorrect wave (for example a slower Rayleigh wave) to measure the time-of-flight. Both plots show reasonable agreement between the measured velocity profiles and the actual profiles. However, there was a large maximum error between the measured and actual velocities of 6.3 % and 17.4 % for the isotropic and anisotropic case respectively.

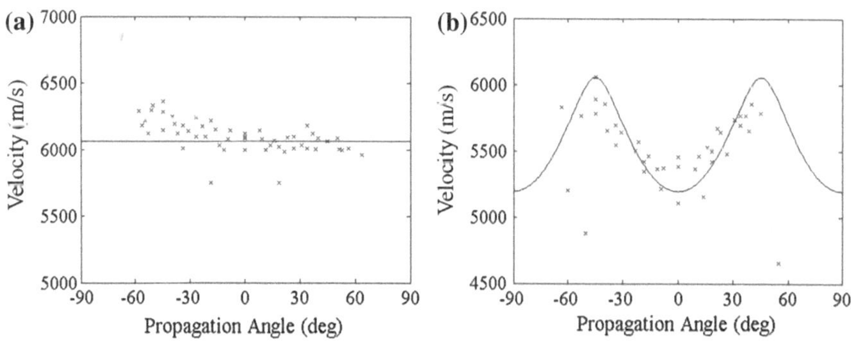

Fig. 5.8 Surface velocity profiles of **a** isotropic aluminium and **b** anisotropic CMSX-4 measured using the SSLW element. The actual velocity profiles are shown as *solid black lines*

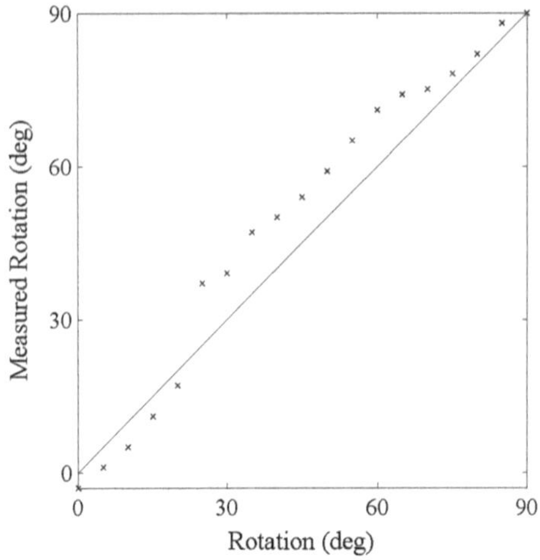

Fig. 5.9 Graph showing the measured orientation using the SSLW element, as the single crystal component was rotated through 90°

To measure the orientation accuracy of the SSLW element, the probe was placed on a single crystal CMSX-4 specimen. The probe was initially aligned to the < 001 > crystallographic direction of the specimen. The specimen was then rotated through 90°. At 5° increments the orientation was measured using the SSLW method.

Figure 5.9 shows the results of this experiment. From examining the results, the maximum error between the true rotation and the measured rotation was 12°. The mean error was found to be 6.5°. This is a much lower accuracy than was achieved with the 3 MHZ 11 × 11 array in Sect. 4.2. However, the 3 MHz array was much larger so the SSLW had time to separate from element ring down. In addition, data from every transmit-receive combination was used and hence the surface velocity could be measured through a full 360°. The engine probe, conversely, only computes the velocity profile from a single element transmitting SSLWs to the array probe elements. Therefore, there is a limited range of propagation angles that the surface velocity profile is measured over.

5.4 Ultrasonic Array Probe Performance

The performance of the in situ engine probe was tested on the actual turbine blades. Some of the blades contained 3 mm long through-going slots located where cracks had been found in the experimental blades, as shown in Fig. 5.10. The effect of the anisotropic velocity profile, beam amplitude variation, crystallographic misorientation and defect size and orientation will be examined in turn. Following this, the orientation accuracy of the probe on actual turbine blades will be examined. This analysis will aid the design of the inspection procedure used for this inspection of blades containing real defects, discussed in the next chapter.

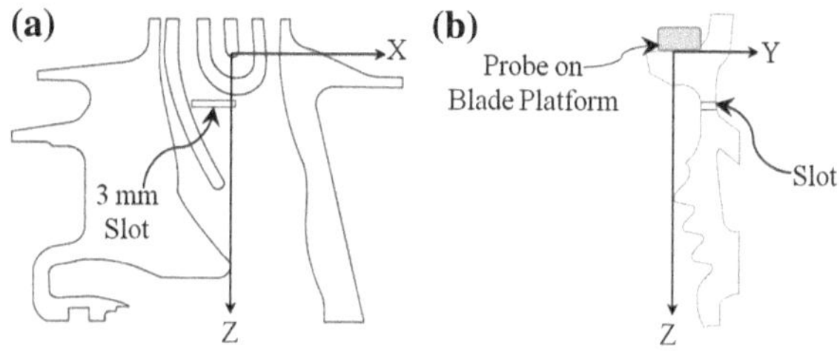

Fig. 5.10 Schematic diagrams showing the in situ engine probe on a turbine blade with a 3 mm long through-going slot. Images courtesy of Rolls-Royce plc

5.4.1 Corrected TFM Images

To demonstrate the effect of the anisotropic velocity profile on the imaging performance of the engine probe, the FMC data from a turbine blade containing a 3 mm long, through-going slot was processed with the TFM algorithm using isotropic and anisotropic velocities. Figure 5.11 shows the uncorrected TFM images of the turbine blade. The image was created assuming a constant velocity of 5.194 km/s (the velocity in the < 001 > crystallographic direction). It can be seen that the anisotropy of the single crystal turbine blade causes significant de-focusing of the image. No features of the turbine blade could be identified with confidence from the images.

Figure 5.12 shows the corrected TFM images of the same blade using the velocity profile from SCORPIO measurements taken from the turbine blade platform. The platform and slot can be easily indentified from these images. The signal response from the slot was 11 dB down from the maximum signal from the blade platform. It was found that the maximum noise level from the same region of a blade without a slot was −23 dB. This gives a 12 dB difference which is a linear signal to noise ratio of 4:1 from the through-going slot.

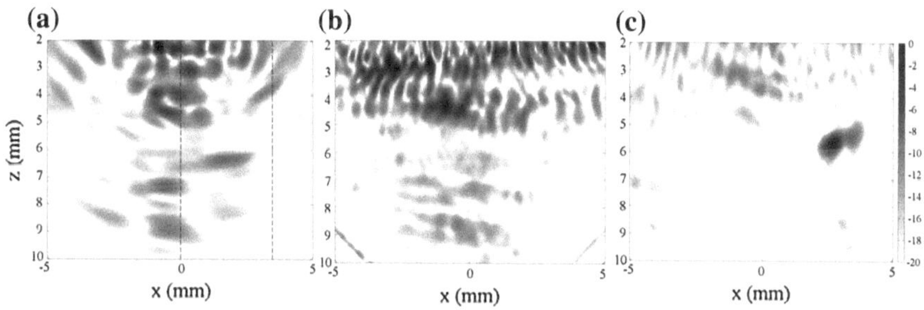

Fig. 5.11 Uncorrected TFM images of a turbine blade containing a 3 mm long through-going slot assuming a constant velocity of 5.194 km/s. The images show **a** the y–z plane at x = 0 mm, **b** the x–z plane at y = 0 mm and **c** the x–z plane at y = 3.5 mm. The images are displayed on a dB scale with 0 dB set to be the maximum signal in image (**a**). The vertical *dashed-lines* in **a** indicate the position in which the image slices in **b** and **c** have been computed

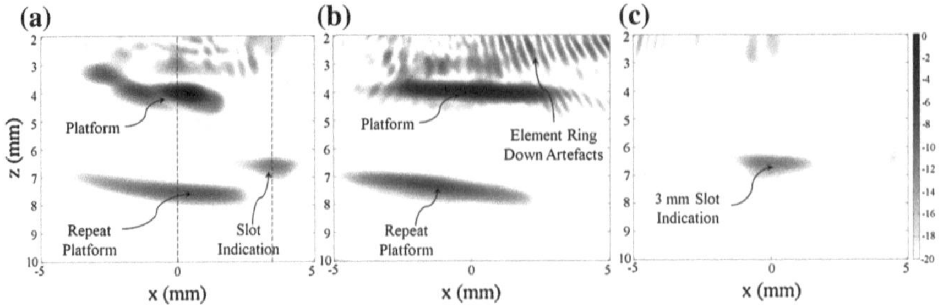

Fig. 5.12 Corrected TFM images of a turbine blade containing a 3 mm long through-going slot. The images show **a** the y–z plane at x = 0 mm, **b** the x–z plane at y = 0 mm and **c** the x–z plane at y = 3.5 mm. The images are displayed on a dB scale with 0 dB set to be the maximum platform signal

5.4.2 *Effect of Beam Amplitude*

Figure 5.13 shows a schematic diagram of the orientation of the crystal in a turbine blade. Angle φ is the maximum deviation of the closest $< 001 >$ crystallographic direction from the primary axis of the turbine blade. For a typical blade, the maximum value φ is permitted to be is 8.5° to maximise the high temperature creep resistance of the single crystal superalloy. The mean value of φ is typically around 5°, whilst there is no control on angle θ, as shown in Fig. 5.13. These two angles can be related to the orientation angles used previously in this thesis (shown in Fig. 4.2) as:

$$\begin{aligned} \varphi &= \operatorname{acos}(\cos(r_X)\cos(r_Y)) \\ \theta &= r_Z \end{aligned} \tag{5.3}$$

where r is the rotation around the axis indicated by the subscript.

To investigate the effect of the beam amplitude at different crystallographic orientations, beam amplitude fields were simulated for the engine probe using the

Fig. 5.13 Orientation of the crystal in a turbine blade. Image courtesy of Rolls-Royce plc

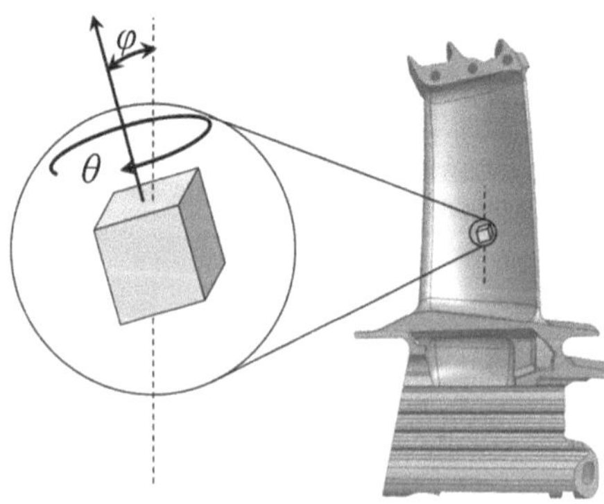

same analysis as detailed in Sect. 3.4. Figure 5.14 shows the TFM response from point-like reflectors at every region within the turbine blade. The images are displayed using the dB scale, where 0 dB has been set to equal the TFM response of the engine probe from a point-reflector assuming the elements were perfect point-like transmitters and receivers. The plots (a) and (b) show the response when the crystallographic orientation φ was 0 and 8.5° respectively. The value 8.5° was chosen as this is the maximum permitted deviation of the $< 001 >$ crystallographic direction from the axis of the turbine blade. By examining the two plots, it can be seen that at $\varphi = 8.5°$ the effect of the anisotropy is to direct ultrasonic energy directly into the shank region and therefore defects in this region should have higher amplitude.

Figure 5.15 shows the maximum TFM response from the defect region (y = 1.5–2.5 mm and z = 6–6.5 mm) at these two orientations. As there is no control of the orientation of the crystal defined by θ in Fig. 5.13 the response is shown at all possible values of this orientation. It can be seen that there is a difference of 17 dB between the maximum response when ultrasonic energy is directed into the shank, as shown in Fig. 5.14b, and the minimum where the energy is directed in the opposite direction to the shank region.

This result may seem to imply that the TFM image intensity should be increased for regions of low ultrasonic amplitude and decreased in high amplitude regions to make the response from reflectors more uniform. However, the ability to detect a defect is due to the signal to noise ratio (SNR) in the local region surrounding a potential defect. A blade with no defect will still have a finite amount

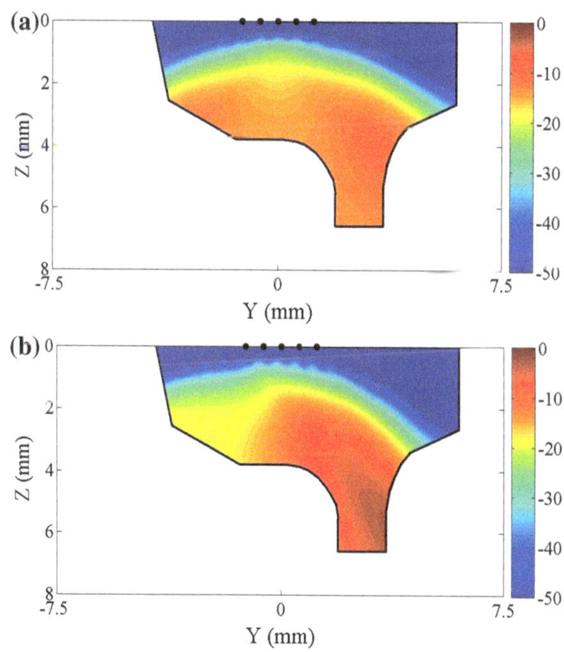

Fig. 5.14 TFM response from point-reflectors at every location in the turbine blade for the engine probe. The orientations of crystal used to simulate the fields are **a** $\varphi = 0°$ and **b** $\varphi = 8.5°$. 0 dB is set to equal the response from perfect point-like transmitters, reflectors and receivers

of noise in the defect region. This is caused by the signals from the blade geometry, and will therefore always be present even if all other causes of noise, for example electrical interference and inter-element cross-talk, are completely eliminated. Therefore, if the TFM image was compensated for the variation in beam amplitude, the SNR of the defect would remain the same, as the TFM intensity from a defect would be increased by the same amount as the noise from a defect-free blade.[1] In conclusion, the beam amplitude compensation would not aid defect detection but it could explain why some defects could potentially be missed if the crystal is in an unfavourable orientation. In addition, it could be used to predict the size of a defect from its amplitude response if other defect variables, apart from defect size and crystallographic orientation, do not change considerably.

Another potential benefit in knowing the beam amplitude due to the anisotropy of single crystals is that array probes could be designed so that the maximum amplitude is sent towards known defect regions. For example, if the whole array probe was moved away from the shank region, in theory, more ultrasonic energy would be sent into the shank. However, in this example, the shoulder of the blade platform would shadow the shank region if the probe was moved further back. In addition, Sect. 5.3.2 showed that, for this geometry, moving the probe backwards would decrease the amplitude of scattered waves, opposing the effect of the increased incident beam amplitude. This may not be the case for other inspections, and therefore the beam amplitude pattern could be used to an advantage.

5.4.3 Effect of Crystallographic Orientation on Defect Detection

To investigate the effect of crystallographic orientation on the defect response, the following experiment was performed. The FMC data was collected from a slotted blade. The TFM algorithm was run on the data using the crystallographic orientation measurements from the X-ray diffraction SCORPIO system. The algorithm was then re-run on the data at every possible orientation.

Figure 5.16 shows the TFM response from the through-going slot at different crystallographic orientations. The plots are shown on a dB scale where 0 dB has been set to the maximum value of noise in blades with no slots. The angle of the platform in the x–z plane is 6°, and in the y–z plane it is zero. The maximum and

[1] To explain this point further, consider depth amplitude correction (DAC) curves used commonly for single element inspections. The beam amplitude decreases with penetration depth in the far-field of the transducer due to beam spread and attenuation. DAC is used to increase the gain of signals from reflectors that occur at greater depths within the specimen so that equivalent reflectors give the same amplitude response. This allows for accurate sizing of small defects. However, if the reflector is too deep, its amplitude response will be similar to the noise level and will be un-detectable, regardless of whether DAC is applied or not.

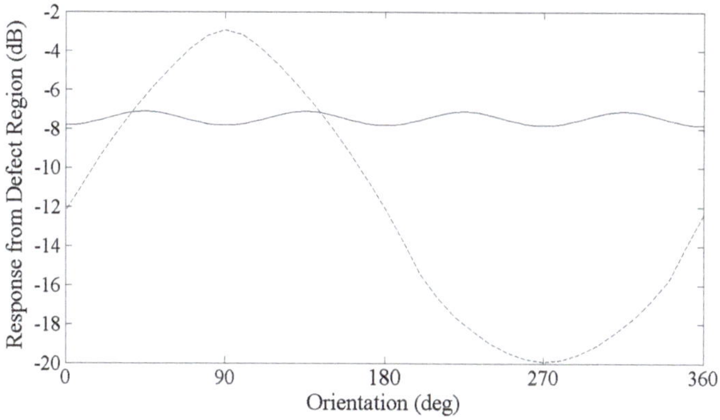

Fig. 5.15 Graph showing the maximum TFM response from point-like reflectors in the defect region of the shank as the crystal rotates around the axis of the blade (θ from 0 to 360°). The *solid line* is for $\theta = 0°$ and the dashed line is for $\varphi = 8.5°$

mean deviations of the crystallographic orientation are shown as dashed circles in Fig. 5.16a. It can be seen that this blade was on the limit of permissible orientation deviation.

It was found that the largest response from the slot was not from the SCORPIO orientation, indicated on the plots as +, but from an orientation that causes an over-sharpening of the defect. At this orientation, indicated with a ×, the defect response is the greatest but the definition of the image was decreased, as shown in Fig. 5.17.

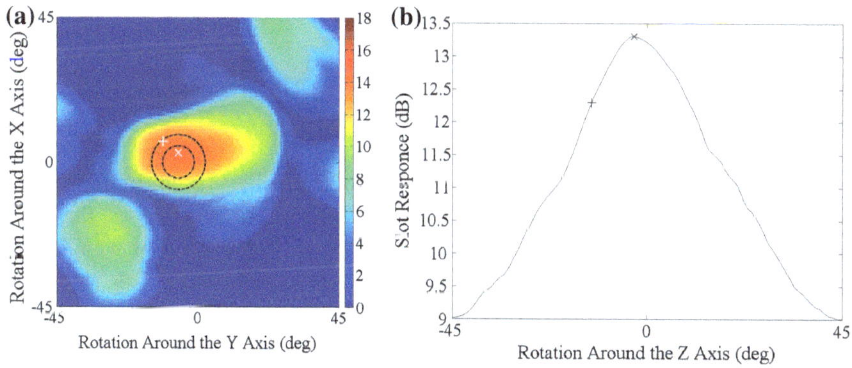

Fig. 5.16 Response from the slot as the crystal orientations used for imaging are changed. The two plots are shown in the plane of the SCORPIO orientation. The plots are shown on a dB scale where 0 dB has been set to the maximum noise level in blades with no slots. The + show the orientation found from SCORPIO measurements and the × show the orientation that gave the maximum signal response. The *outer circle* in **a** shows the maximum permissible orientation of the blade and the *inner circle* shows the mean orientation that the blade may be at

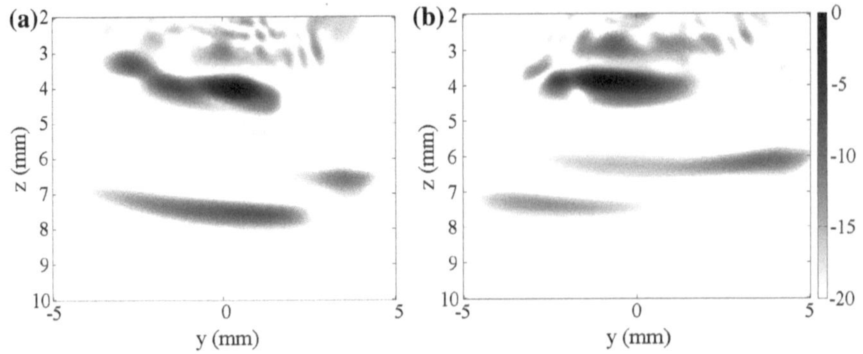

Fig. 5.17 TFM images of the blade computed using **a** the SCORPIO orientation and **b** the orientation that gave the highest slot response. Notice how the chamfer of the platform has much better definition in (**a**) than in (**b**)

The slot response from the SCORPIO measurement, the maximum response and the response that assumes $\theta = 0°$ (i.e. the centre of the circles in Fig. 5.16a are shown in Table 5.1. It can be seen that there is a minimal reduction in slot amplitude, assuming $\theta = 0°$, from the maximum value. This result therefore suggests that by simply assuming $\theta = 0°$ when processing the TFM algorithm, the effect of defect detection due to crystallographic misorientation is minimal. The deviation of the orientation of most turbine blades is less than this example, and therefore there would be even less effect on the slot response if $\theta = 0°$ was assumed for the majority of turbine blades. In other words, it is not strictly necessary to measure the crystallographic orientation, as the manufacturing tolerances do not allow the orientation to deviate to such an extent that it has a negative effect on defect detection.

5.4.4 *Effect of Defect Size and Orientation*

In all the examples so far, flat, through-going slots have been used. In this section the effect of slot geometry (through-going extent and slot angle) on the TFM response is investigated. This was done using isotropic test pieces that replicated the turbine blade cross-section, with different slot geometries, as shown in

Table 5.1 The slot response found using the SCORPIO measurements, the maximum response and response assuming $\theta = 0°$. 0 dB has been set to the maximum noise level in blades with no slots

Orientation	Response from slot (dB)
SCORPIO	14.1
Maximum	14.7
Zero deviation	14.5

Fig. 5.18. Isotropic test pieces were used for two reasons: firstly the effect of anisotropy is removed and therefore the effect of defect geometry can be examined in isolation. Secondly, the slots are created using electrical discharge machining (EDM). The geometry of the single crystal turbine blade was found to interfere with the EDM tooling when the angled slots were being machined.

Figure 5.19 shows a graph of slot response as slot angle was varied for through-going slots from −40° to 40°. The maximum level of noise in the slot region for a specimen with no slot has been set to 0 dB. It can be seen from the graph that the slot response reached a maximum at a slot angle of 10°. At this angle the ultrasonic energy from the array is almost normal to the surface of the slot, and therefore the majority of energy is reflected back towards the array. At other angles, the incident waves are oblique to the slot surface and a much smaller proportion of the ultrasonic energy is reflected towards the array. The difference between the maximum and minimum for these through-going slots was 12.3 dB.

Figure 5.20 shows a graph of slot response as the through-going extent of slots at 0° and 30° was varied from 10 to 90 % of the shank wall. In general, the response increases with extent as there is more slot area to reflect ultrasonic energy. Beyond 50 % the response begins to plateau. This is because the shoulder of the platform shadows the shank region into which the slot extends. In fact the response decreases for the 0° slot beyond 50 % through-going extent. However, a reason for this is not clear.[2]

The strong variation of the response to slot angle and the plateau effect for slots greater than 50 % shank wall thickness mean that the slots cannot be sized, in terms of the through-going extent, on signal response. In addition, the extent of the slot indication in the TFM image in the y–z plane did not change, meaning the slot extent could not be measured directly from the image either.

5.4.5 Crystallographic Orientation Accuracy

To measure the orientation measurement accuracy of the in situ engine probe, the crystallographic orientation of a HPT1 blade was first measured from the blade platform using the SCORPIO system. The SSLW and image-based ultrasonic orientation methods were then used in turn to predict the orientation.

[2] The experiment on the 0° slots with different through-going extents was repeated numerous times. Every experiment showed that the 50% through-going slot gave a significantly larger response over the 80 and 90% slots. A 3D X-ray CT scan was used to show that the slot angle, depth and through-going extent were correct. Replicas of the slots were taken and showed the roughness of the slots were very similar. Simple 2D FE models were run to show that the slot response, for 0° slots with increasing through-going extent, increased up to 60% through-going extent and then plateau. Therefore, the cause of this drop off in signal at 80 and 90% is not fully understood. Possibly, the effect is due to interference from the 3D scattering off the tips of the slots. However, this could only be verified from full 3D FE simulations, which are not currently possible on computers available for this project.

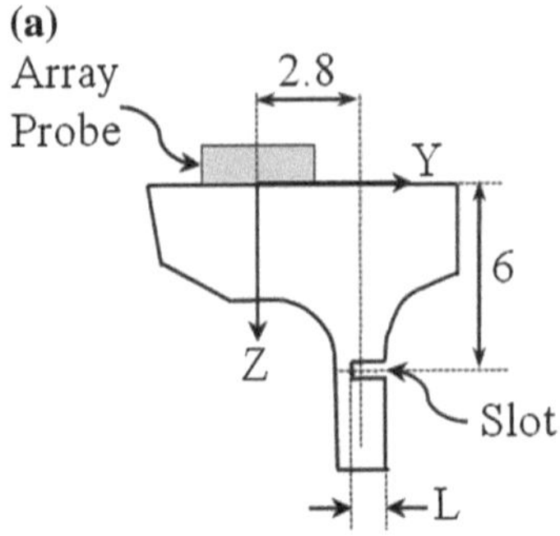

Fig. 5.18 Schematic diagram showing isotropic test pieces with **a** different slot through-going extents, L, and **b** different slot angles, θ. All dimensions are shown in mm

It was found that the SSLW could not be extracted from the time-traces. It was thought that as the SSLW element was close to the edge of the blade platform the reflected bulk and surface waves interfered with the SSLW signals.

The image-based method was performed by collecting the FMC data and then computing the TFM image using the SCORPIO orientation measurements. Slices of this TFM image were used as the reference images. The probe was then replaced on the same turbine blades and the FMC data was collected. The TFM images for this data were processed at all possible orientations, and at each iteration the images were cross-correlated against the reference images. Figure 5.21 shows the plot of cross-correlation value against orientation. The $\times$ indicates the location of highest correlation and the $+$ is the SCORPIO orientation used to

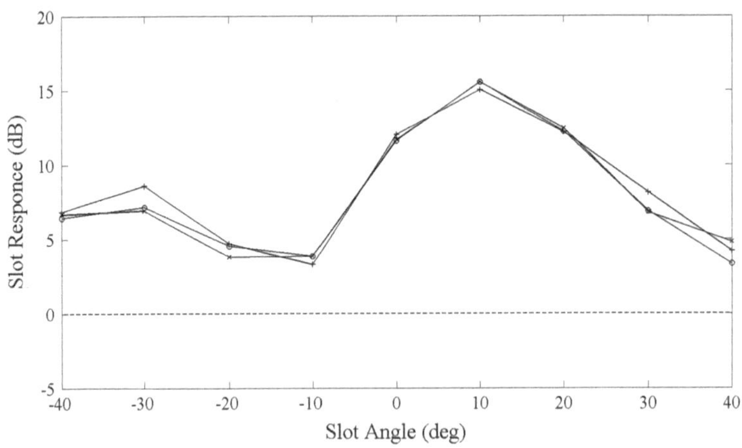

Fig. 5.19 Graph showing slot response as slot angle was varied for though-going slots. 0 dB is set to the maximum level of noise in the slot region for a specimen with no slot. The three lines show repeat measurements for the same slot

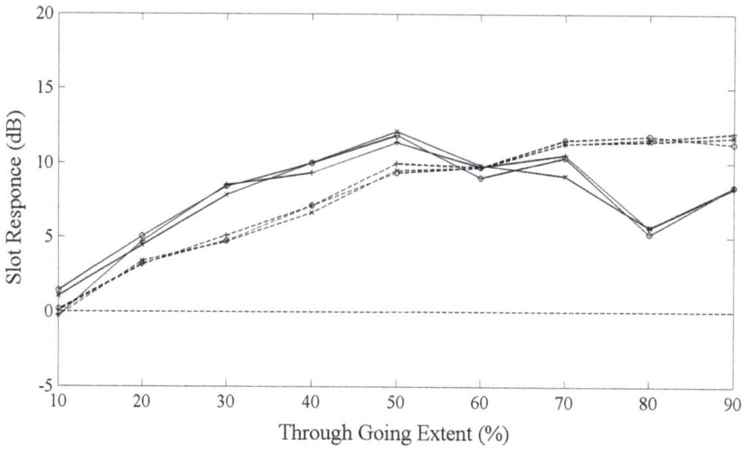

Fig. 5.20 Graph showing slot response as the through-going extent was varied for 0° (*solid line*) and 30° (*dashed line*) slots. 0 dB is set to the maximum level of noise in the slot region for a specimen with no slot. The three lines show repeat measurements for the same slot

compute the reference image. The orientations of these two points are shown in Table 5.2.

The results shows that even by cross-correlating the images of the same blade but using two sets of data, the predicted orientation differs from the SCORPIO results by up to 5°. Clearly, if the same FMC data was cross-correlated against itself the correlation value at the SCORPIO orientation would equal unity, by definition. It was thought that because the probe is small, very slight changes in the

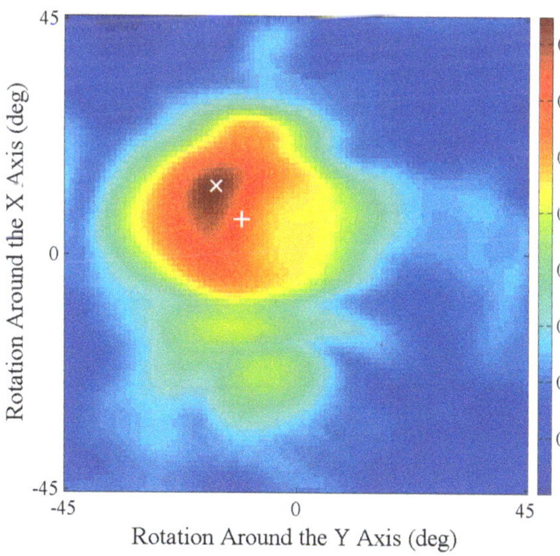

Fig. 5.21 The cross-correlation values between the TFM images computed at different orientations and the reference images from the same blade computed using SCORPIO orientation measurements. The × indicates the location of highest cross-correlation and the + is the SCORPIO orientation

Table 5.2 The orientation measurements found using the SCORPIO X-ray diffraction method and the imaged-based method

Orientation	Rotation around X axis (deg)	Rotation around Y axis (deg)
SCORPIO	6.6	−10.9
Image-based	13	−16

attitude of the probe relative to the blade, when the two sets of data were collected, caused this inaccuracy.

5.5 Discussion

This section has described the design and development of a 2D ultrasonic array probe for the in situ detection of root cracking in stage-1 high-pressure turbine blades. The sensitivity of this probe for the detection of cracks (represented by slots in this section) has been shown to be affected by a number of variables. These variables and their effect on the slot response and signal to noise ratio (SNR) are summarised in Table 5.3.

It was shown that an inspection could not be carried out if the correction for anisotropic velocity profile was not performed. However, it was shown that assuming that the $< 001 >$ crystallographic direction was aligned with the blade axis, instead of using the true orientation, had very little effect on the response from a slot. This was fortunate, as neither of the developed orientation methods could measure the crystallographic orientation to an accuracy of greater than 5°.

The beam amplitude due to the anisotropy of single crystals was also shown to have a large effect on the amplitude response of reflectors. However, any compensation for beam amplitude would increase the response from a defect by the same amount as the noise of a defect-free blade, and therefore there is little point in performing this compensation.

The angle and through-going extent of the slots were also shown to have a large effect on the TFM response. This means that the defects cannot be accurately sized, in terms of through-going extent, as a large defect at an unfavourable angle would give the same response as a small defect at a favourable angle.

Table 5.3 The effect of different variables on slot response and signal to noise ratio (SNR)

Variable	Effect on slot response (dB)	SNR (dB)
Velocity correction	Vital	Vital
Beam amplitude correction	17	0
Crystallographic misalignment	0.4	0.4
Effect of through-going slot angle	12.3	12.3
Effect of through-going extent	12.1	12.1

References

1. Drinkwater BW, Wilcox PD (2006) Ultrasonic arrays for non-destructive evaluation: a review. NDT E Int 39(7):525–541
2. Velichko A, Wilcox PD (2010) Strategies for ultrasound imaging using two-dimensional arrays. Rev Prog Quant Nondestr Eval 29:887–894

Chapter 6
The Assessment of the Developed Inspection Capability

Abstract In the previous chapter, the development of an ultrasonic array system for the inspection of a specific single-crystal turbine blade was presented. The inspection was designed to be performed in situ and detect cracks which initiate from the internal surface of the blade root. The performance of the system was evaluated using artificial defects. This chapter evaluates the inspection capability of the developed array system using turbine blades containing real cracks. The evaluation is performed using probability of detection (PoD) analysis. This is a statistical method that is used widely to evaluate NDE processes in a rigorous manner. A PoD analysis provides a quantitative measure of the probability of detecting a defect of any given size. To perform the PoD analysis for the array system, the ultrasonic response was recorded from cracked blades installed inside an engine mock-up which represented the in-service inspection conditions as accurately as possible. This process was also performed using a conventional single-element probe. After inspection, the cracked turbine blades were cut open to measure the geometry of the cracks. The results of the PoD analysis for the two ultrasonic systems are compared and the array system is shown to offer significant improvements in crack detection capability. However, it is shown that the developed array system suffers from poor reliability. The factors affecting the inspection reliability are discussed.

6.1 Introduction

In this chapter, the defect detection capability of the developed inspection system will be analysed in a quantitative manner. The crack parameter that is of greatest interest is the through-going extent of the crack in the shank wall. This is the parameter that most accurately determines the structural integrity and remaining safe operational life of the turbine blade. Therefore, the response, in terms of size

C. Lane, *The Development of a 2D Ultrasonic Array Inspection for Single Crystal Turbine Blades*, Springer Theses, DOI: 10.1007/978-3-319-02517-9_6,

and amplitude, of the crack indication in the TFM image would ideally increase significantly with through-going extent. However, experiments in the previous chapter with slots showed that this is unlikely, as other variables, such as crack angle, have an equally significant effect.

In this chapter, actual cracks in turbine blades will be investigated. The blades came from five demonstrator engines that were run to a very high number of cycles and therefore were likely to contain defects. Figure 6.1 shows example photographs of cracks in the shank wall of the turbine blades. Figure 6.1a is a view in the *y*–*z* plane (referring to Fig. 5.10) that has been created by sectioning the blade from root to tip, then polishing the blade back until the maximum extent of the crack is revealed. This process can provide crack length and crack angle measurements but is very time-consuming. Figure 6.1b is a view in the *x*–*y* plane created by notching the blade in the shank wall region and then forcing the crack open. This process is very rapid but can only provide a measurement of the crack length, and sometimes does not open the actual crack face.

The five blade sets were inspected with the ultrasonic system on a mock-up engine, which had the exact geometry of the real demonstrator engine, but allowed direct access to the back of the stage-1 turbine so the blade sets could be changed over quickly. The ultrasonic system was operated without beam amplitude correction and used a velocity profile that assumed that the <001> crystallographic direction was aligned with the axis of the blade. Each engine set, containing 74 blades, took approximately 10 hours to inspect.

After the blades were ultrasonically inspected, the defects were forced open to measure the through-going extent of each defect. Ideally the full geometry of each defect would have been measured, either by sectioning and polishing, or by using X-ray computed tomography (CT) [1]. In addition, it would have been useful to have the crystallographic orientation of each blade measured by SCORPIO. However, considering that 370 blades were inspected, the resource to measure the full turbine blade and crack geometry or the orientation was not available for this project. Furthermore, there are a large number of inspection variables that could

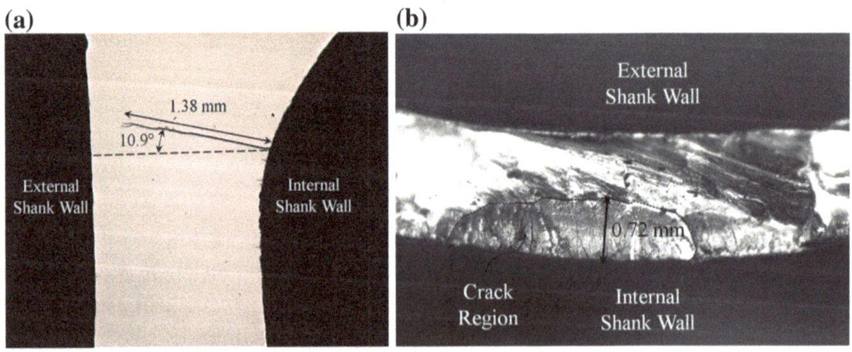

Fig. 6.1 Photographs of cracks in the shank wall of the experimental HPT1 blade. Images courtesy of Rolls-Royce plc

also affect ultrasonic response including: crack roughness, depth, length, area and orientation in the x–z plane and in the y–z plane, multiple cracks and variation in the turbine blade geometry. Therefore, it would be far too onerous to investigate the effect of each individual variable on the ultrasonic response. Instead, a probability of detection (PoD) analysis will be performed. This is a statistical method used frequently in NDE [2–6] which takes into account all of the inspection variables and predicts the probability of detecting a defect of a given size.

6.2 Raw Inspection Results

Prior to inspecting the run engine sets, 10 defect-free blades were inspected to measure the noise baseline. For the inspection of the engine sets, 0 dB was set to equal the maximum noise measured from the 10 defect-free blades. Two calibration blades were inspected before and after the inspection of the engine sets: one with a through-going slot and one that was defect-free, to show that the inspection system was consistent.

Figure 6.2 shows a graph of ultrasonic signal response against the through-going extent of the cracks in the shank wall of the turbine blade measured after forcing open the cracks. A slight increase of ultrasonic response with through-going extent can just about be discerned. However, there is a large amount of variation in the ultrasonic response for a given crack size. For example, there is a difference of 8 dB for 100 % through-going cracks. Two of these through-going cracks gave signals below the noise level of the defect-free turbine blades. The large spread of response can be explained by the number of inspection variables that have been discussed previously. To illustrate, upon cutting open the turbine

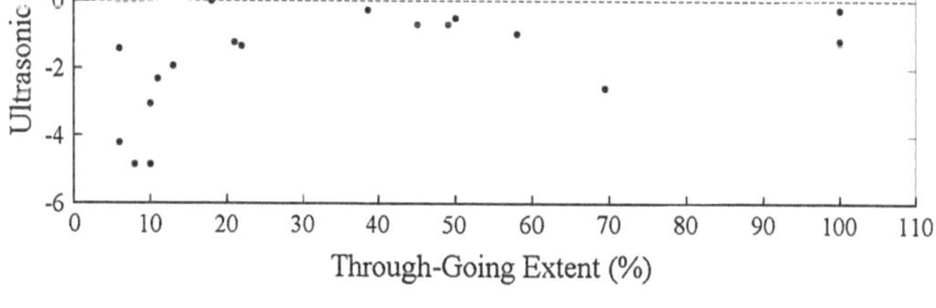

Fig. 6.2 Graph showing the relationship between the ultrasonic response and through-going extent from the run engine set turbine blades. 0 dB has been set to the maximum level of noise recorded in the defect-free blades

blades, it was found that the average shank wall thickness from all the blades was 1.50 mm. However, this varied between 1.30 and 1.76 mm. This change in wall thickness would alter the distance between the array probe and the shank wall region, also affecting the ultrasonic response from a defect.

It is worth noting that the ultrasonic response from the through-going slot, described in the previous chapter, was 12 dB above the maximum level of noise. This is, therefore, 5 dB higher than the maximum response from the through-going cracks.

Figure 6.3a, b shows example TFM images of real cracks that gave a similar ultrasonic response, but had a through-going extent of 21 and 100 % respectively. It can be seen that the ultrasonic amplitude and the overall size of the indications from the two cracks are similar. Therefore, the system was unable to size the cracks.

6.3 Probability of Detection Results

In this section, probability of detection (PoD) analysis is applied to the raw data. The aim is to provide a quantitative measure of the probability of detecting a crack of a given size by taking into account the many variables that may affect the crack signal. The background theory of PoD and how variations are included in the analysis is presented in Appendix E.

The first step is to apply a model that predicts the response of a crack for a given size. The models that were considered are shown in Fig. 6.4. The models took the form:

$$z = \beta_1 + \beta_2 x \tag{6.1}$$

were z is the predicted response and x is the crack size. β_1 and β_2 are coefficients found by comparing the modelled data, z, and real data, y, by maximising the expression:

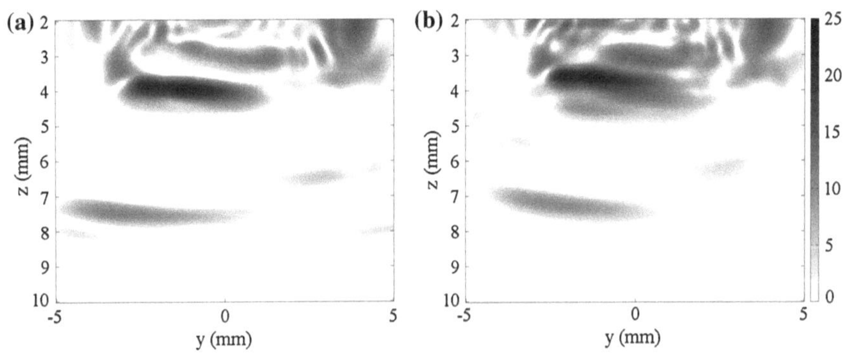

Fig. 6.3 Example TFM images showing cracks that are **a** 21 % and **b** 100 % through-going

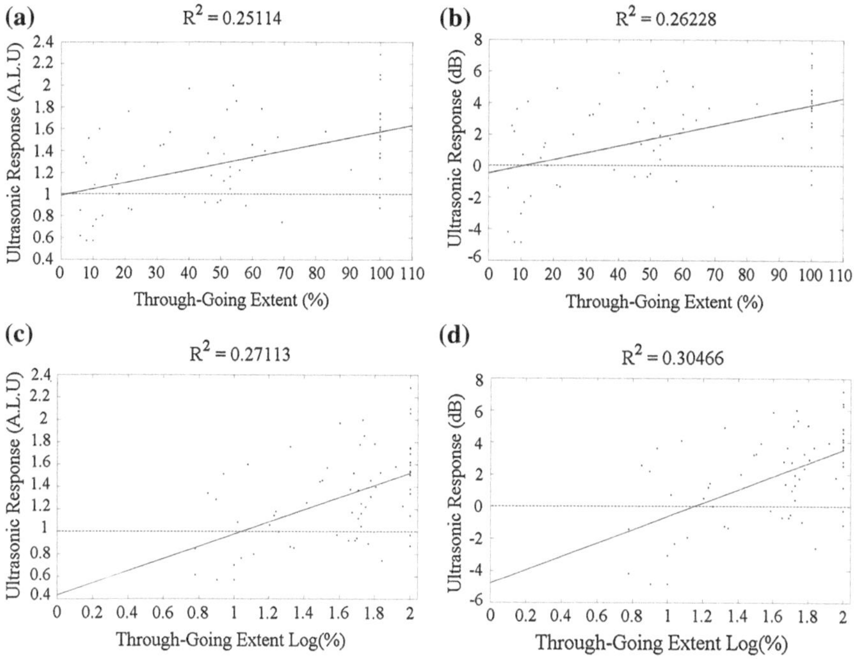

Fig. 6.4 Graphs showing the fit between the actual data and **a** linear–linear, **b** linear–log, **c** log–linear and **d** log–log models. The *dashed line* in each graph shows the maximum level of noise from defect-free blades

$$R^2 = 1 - \frac{\sum_i^n (y_i - z_i)^2}{\sum_i^n (y_i - \bar{y}_i)^2} \tag{6.2}$$

where the bar indicates the mean and n is the number of cracks in the data set. When R^2 is unity there is a perfect fit between the modelled and real data. The variables z and x can be on either a linear scale or a log scale. Therefore, the four permutations of the models are shown in Fig. 6.4. On graphs a and c, A.L.U stands for arbitrary linear units, where data has been normalised so the maximum level of noise equals one.

Figure 6.4 shows that the best fitting model (the model that gave the highest R^2 value) used the log–log scale. Therefore, this model was used to compute the PoD curves. It should be noted that all models gave a relatively low R^2 value, as the models were not particularly suited to predicting the ultrasonic response based on the through-going extent alone. As stated previously, this was due to the large effect of the other inspection variables on the ultrasonic response.

The next step towards computing the PoD curve is to choose a signal threshold. If the threshold is set too low, then signals from defect-free blades will be higher than this threshold, resulting in many false calls. Conversely, if the threshold is too high, blades with large cracks will give responses less than the threshold, and the probability of detecting these defects will be very low. Figure 6.5 shows the PoD curve that has been computed using a threshold that is 1.5 times greater than the maximum noise level from defect-free blades. The PoD curve shows that the probability of detecting 100 % through-going cracks is 50 %. The PoD curve shown in Fig. 6.5 and all other PoD values have all been computed with a 50 % confidence level (see Appendix E).

Table 6.1 shows the probability of detecting 100 % through-going cracks for a given threshold. The percentage of through-going cracks from the real data that would have been classed as defected using each threshold is also shown. The results show that as the threshold is increased, the PoD reduces.

In the aerospace industry, PoD results typically state the smallest size of defect that can be found with a PoD of 90 % and a confidence level of 95 %. However, this value cannot be given for the developed array system because a 90 % PoD level was not achieved for any defect size.

6.4 Capability of the Single Element Probe

By way of comparison, the single element probe (described in Appendix D) was used to inspect the 18 through-going cracks prior to the cracks being forced open. The gain of the single element probe was set so that the signal response from a

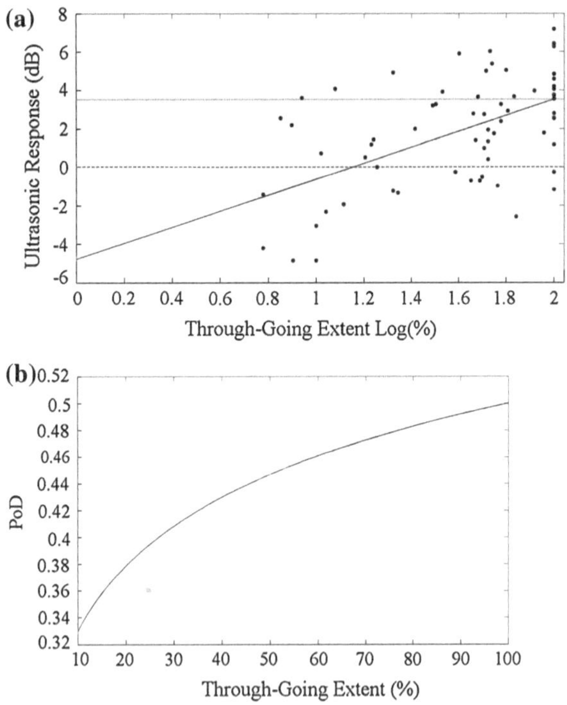

Fig. 6.5 Graphs showing **a** the raw data on a log–log scale with a signal threshold (*dotted line*) of 1.5 × the noise level and **b** the PoD curve for these parameters

Table 6.1 The PoD of 100 % through-going cracks for a given threshold and the percentage of through-going cracks greater than the threshold

Threshold (× noise level)	100 % Through-going PoD	Number of through-going cracks > threshold (%)
1.00	64.6	88.9
1.25	56.7	83.3
1.50	50.0	66.7
1.75	44.3	16.7
2.00	39.5	16.7

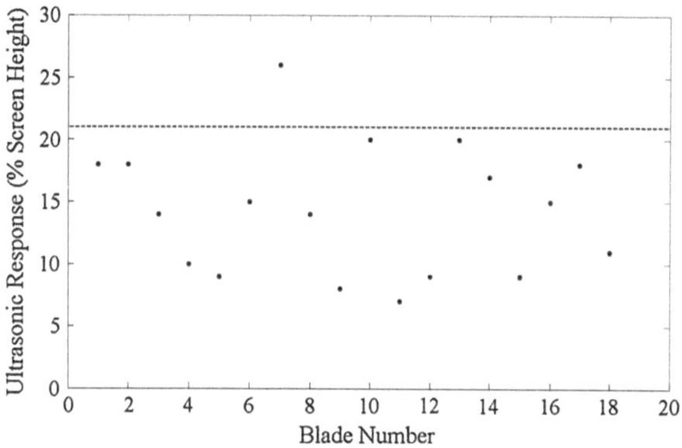

Fig. 6.6 Ultrasonic response of the single element probe from 100 % through-going cracks

calibration blade with a 3 mm long through-going slot was 100 % of the screen height of the ultrasonic A-scan. At this gain setting, the maximum noise level from defect-free blades was 21 % screen height.

Figure 6.6 shows the ultrasonic response from the 100 % through-going cracks with the single element probe. It can be seen that only one of the cracks gives an ultrasonic signal greater than the noise threshold. By comparison, the array system detected 16 out of the 18 through-going cracks giving a defect signal greater than the noise threshold. Therefore, the developed array system offers significant defect detection capability over the single element probe. Furthermore, the results from the single element probe were so poor there was no point in performing a full PoD evaluation.

6.5 Concluding Remarks

This section has shown that there is a very large variation in the signal response, with the ultrasonic array inspection developed by this project, from cracks in the shank wall of the HPT1 blades. The cause of this variation has been discussed.

PoD analysis has been used to predict the probability of detecting a crack of a given size. It has been shown that the probability of detecting 100 % through-going cracks using the array system is relatively low. For example, using a detection threshold that is 1.5 times the maximum noise level from defect-free blades, only 50 % of the through-going cracks would be detected. However, the single element probe could only detect one out of 18 through-going cracks using a threshold set to equal the maximum noise level from defect-free blades. The array system therefore offers a significant improvement in detection sensitivity over a conventional single element probe. Neither system was able to offer any capability in sizing the cracks.

References

1. Brunke O (2010) High-resolution CT-based defect analysis and dimensional measurement. Insight 52(2):91–93
2. Georgiou GA (2006) Probability of detection (PoD) curves: derivation, applications and limitations. Research report 454, Health & Safety Executive Books, Suffolk
3. Nondestructive Evaluation System Reliability Assessment (1999) MIL-HDBK-1823. USA Department of Defence Handbook
4. Schneider CRA, Rudlin JR (2004) Review of statistical methods used in quantifying NDT reliability. Insight 46(2):77–79
5. Thompson RB (2005) Use of physics-based models in the determination of POD for 3D Defects. Rev Prog Quant Nondestr Eval 24:1917–1924
6. Annis C (2011) Statistical engineering. http://www.statisticalengineering.com. Accessed 23 April 2011

Chapter 7
Conclusion

Abstract This thesis has described the development of an ultrasonic array system for the in situ inspection of single-crystal turbine blades. The system was intended to detect and characterise defects in the turbine blades with the highest possible accuracy to mitigate the risk of in-service failure. Single-crystal turbine blades are highly anisotropic. This has been shown to deleteriously affect the inspection of these components with ultrasonic NDE methods. Therefore, the earlier chapters of this thesis developed methods to measure and compensate for the anisotropic behaviour of single-crystal components to ensure the most accurate inspection possible. The later chapters detailed the development of an array system for the inspection of a specific turbine blade to detect root cracking. The inspection performance of the developed system was evaluated using artificial and real cracks. These evaluations showed the array system to offer greatly improved crack detection over a conventional ultrasonic inspection. However, the system was shown to exhibit lower detection sensitivity than would be required for a viable in-service inspection for this particular turbine blade scenario. In this conclusion chapter, the main findings of the thesis are presented in detail. In addition, further research work is suggested that could be used to improve the detection sensitivity and lead to accurate defect sizing for future inspection systems.

7.1 Findings of Thesis

The aim of this thesis was to develop an ultrasonic array system for the in situ inspection of single crystal turbine blades for cracking in the root section of the blade. The defect detection sensitivity was to be as high as possible, and the detection and characterisation capability of the developed system was to be evaluated in a quantitative manner.

There were two main motivations behind this project, as discussed in the introduction in Chap. 1. The first was that jet-engines are increasingly sold with *"power-by-the-hour"* contracts, which results in the engine manufacturer being

C. Lane, *The Development of a 2D Ultrasonic Array Inspection for Single Crystal Turbine Blades*, Springer Theses, DOI: 10.1007/978-3-319-02517-9_7,

responsible for the maintenance of the engine. For these contracts to be competitive and profitable, the engine components should operate for as long as possible whilst meeting strict safety requirements. One way to achieve this is by periodically inspecting engine components for defects that may have initiated in-service. The ability of these inspection systems to be used in situ, removes the time-consuming and extremely expensive cost of removing the components from the engine prior to inspection.

The second project driver was the increasing use of the damage-tolerant philosophy in the design of new aerospace components. Damage-tolerance design assumes that all components contain defects. The size of the defects is determined by the capability of the inspection system used prior to the component entering service. It is foreseeable that, as this design philosophy becomes more widespread, and the experience of operating with *known* defects increases, a component with a defect that has been detected in-service may be able to operate safely before it is later replaced. This is not to suggest that jet-engines would be permitted to fly for a great number of cycles before the defective component is replaced. Rather, if the defect can be measured to a high accuracy, the engine may be used to fly the aircraft to an engine maintenance centre where the engine servicing can be performed. This would therefore require in situ inspections to have high defect characterisation capability, as well as high detection sensitivity, and a quantitative measure of the accuracy and reliability of both.

The main difficulty faced in this project was to understand and compensate for the anisotropic behaviour of single crystals. The theory and models of wave propagation in anisotropic materials required for this project were presented in Chap. 2. The models can be easily applied to any form of elastic anisotropy. The only limitation is that the analytical model for beam amplitude is not valid for slowness surfaces with zero curvature and cusps, for example the qS_V mode of single crystal nickel-based alloys. However, the numerical finite-element model could cope with this type of anisotropy.

The complications of inspecting anisotropic single crystal components were presented in this project in three ways, namely: the variation in velocity with propagation direction, the beam directivities from ultrasonic transducers in anisotropic media, and the variation in crystallographic orientation from component to component.

In Chap. 3, it was shown that by being able to model the velocity variation, ultrasonic array imaging algorithms, such as the TFM algorithm, can be adapted to correct for the effects of anisotropy. It was also shown in this chapter that the beam amplitude patterns in anisotropic media from ultrasonic transducers would affect the amplitude of reflectors, depending on the position of the reflector relative to the array. This is true of both isotropic and anisotropic materials. However, the effect in anisotropic materials is more complex. For example, the maximum amplitude response from an ultrasonic array using longitudinal waves in isotropic media to an omni-directional reflector is always directly below the centre of the array. In anisotropic materials there may be multiple maxima and these could occur at oblique angles to the array. It was shown that this could affect the detectability of a defect. However, the effect could not be compensated for, as all this would do is increase the amplitude from a potential defect by the same amount as the noise in the location of the defect, and hence the signal to noise ratio would not change. The effect of the beam directivity on defect characterisation is discussed in Sect. 7.2.3.

The usefulness of an ultrasonic array probe in measuring the crystallographic orientation of a component as well as being able to perform defect inspections was discussed in Chap. 4. A number of ultrasonic array orientation methods were successfully developed and their accuracy was determined.

In Chap. 5, the work from the previous chapters was used to design a specific ultrasonic array system for the inspection of root cracking in an experimental stage-1 high-pressure turbine blade. It was shown that the correction for velocity variation was vital for the inspection of this component. However, it was found that it was sufficient to assume the [001] crystallographic direction was aligned to the axis of the blade, and therefore the orientation methods were not required. This was fortunate as the orientation methods, for this specific component, were not particularly accurate. This was primarily due to the limitations in the probe design caused by access restrictions and a limited inspection area. For other inspections, the crystallographic orientation may be necessary and the methods developed in this thesis could potentially be successfully implemented.

Furthermore in Chap. 5, the effect of inspection parameters was shown to greatly influence the signal response and hence the detectability of defects. These include: the distance between the probe and the shank wall of the turbine blade, the crack depth, length, through-going extent, orientation, roughness and the crystallographic orientation of the turbine blade.

The capability of the developed ultrasonic array system was evaluated in Chap. 6, using 370 blades that potentially contained cracks. The defect parameter of interest was the through-going extent of the cracks, as this most accurately determined the structural integrity of the turbine blade. It was shown that the system was unable to size defects, as defects with significantly different through-going extents could give the same ultrasonic response in terms of image amplitude and size. In general, the signal response of the cracks increased with through-going extent. However, there was a large amount of scatter within the data caused by the effect of other inspection parameters.

Probability of detection analysis was applied to inspection data. This showed that there was a 50 % probability of detecting a through-going crack when using a detection threshold that would minimise the number of false calls. A more conventional single element probe was shown to be able to only detect one out of 18 through-going cracks using the lowest detection threshold possible.

In conclusion, this project has developed an in situ ultrasonic array system that is capable of detecting defects in single crystal turbine blades with a known reliability. In this respect, the aims of the project have been met and the project has been a success. The low probability of detection and the inability to size defects means that this particular system is unlikely to be deployed to aid in improving the life of the specific turbine blade described in Sects. 5 and 6. However, the work presented in this thesis could be successfully applied to other turbine blade geometries and indeed other single crystal components.

7.2 Suggested Future Work

The findings of this thesis have raised a number of issues and potential avenues for future work. These will now be discussed in turn.

7.2.1 3D Arrays

Fundamentally, the low defect detection rate of the array system developed for this project is due to the distance between the array probe and the defect region. This distance could not be reduced as there was limited amount of space on the flat platform on which the array could be placed. However, if the array probe could be manufactured to be non-flat, i.e. with a 3D profile, the whole array could be moved forward and the front elements could conform to the curvature of the turbine blade aerofoil. This could potentially improve the detection capability of the system. Such arrays are currently being developed by the University of Strathclyde [1].

7.2.2 Ultrasonic Crystallographic Orientation Methods

The majority of single crystal components are inspected with SCORPIO-type X-ray systems to measure their orientation. However, there is a desire to implement alternative orientation systems to eliminate the health and safety issues surrounding the use of ionising radiation. Indeed, this is the express purpose of a PhD project that is currently underway at the University of Nottingham [2]. This system uses ultrasonic surface wave velocity profiles created and detected using lasers to measure crystallographic orientation. However, lasers pose their own safety risk and, therefore, orientation methods using piezoelectric transducers could offer benefits.

7.2.3 Improved Defect Characterisation in Anisotropic Materials using Large Aperture Arrays

The use of FMC has allowed the recent development of novel defect characterisation methods. For large defects, i.e. greater than the ultrasonic wave length, the defect can be sized directly from the TFM image. However, defects in the order of a wavelength are more accurately sized from their amplitude response. The defects can be characterised further by using methods such as the vector-TFM (VTFM) [3]. The VTFM divides the array into sub-apertures, which can be used to compute the reflection amplitude of a defect at different directions. This allows specular defects to be distinguished from omni-directional defects and the orientation of specular defects to be predicted. The extraction of the scattering matrix from the FMC data can produce similar measurements for sub-wavelength defects [4]. The scattering matrix measures the far-field amplitude of a reflector at a range of incident and reflected angles using the individual elements in the array. None of these methods could be implemented using the system developed for this project, as the array aperture was too small. For inspections with larger apertures, these methods could potentially be used for the improved characterisation of defects in anisotropic components. This requires the beam directivity of the array elements to be computed, as the methods are all dependent on amplitude measurements. The work on beam amplitude in anisotropic materials presented in this thesis could be used as a starting point to investigate these characterisation methods.

7.2.4 FMC Data Simulations of Complex Defects

The design of the array element layout for this project was conducted in a rather un-scientific way. Ideally, the FMC data for complex defects and geometries, such as the turbine blade root cracking, would be simulated and this would aid the design and optimisation of the array layout. However, a full 3D numerical finite element simulation would be too large to run on modern computers. For example, the 3D FE model of the turbine blade would be approximately 10 $\times$ 10 $\times$ 5 mm. Assuming a frequency of 10 MHz, a slowest ultrasonic wave velocity of 3 km/s and using 40 elements per wavelength would require 1.2 billion nodes. This is a far greater processing demand than current computers can handle. Ray-tracing models on the other hand can only cope with a limited number of reflections and mode conversions from relatively simple boundaries. However, researchers at the University of Bristol [5] and Imperial College London [6] are developing hybrid models that combine the speed of ray-tracing with the robustness of FE to simulate data for complex defects. Only a minimal amount of work would be required to adapt these models for simulating anisotropic components.

7.2.5 Alternative Imaging Algorithms

A number of new imaging algorithms have been developed recently that use FMC data. For example, researchers at Imperial College London have developed super-resolution imaging algorithms [7]. These algorithms may offer improved defect detection and characterisation but are currently only capable of inspecting isotropic media. Further work is required to investigate whether they could be adapted for the inspection of anisotropic materials, and indeed if they would offer improved defect detection sensitivity for inspections like the one detailed in this thesis.

References

1. Harvey G, Gachagan A, Mackersie JW (2006) Enhanced flexibility and performance from a Piezocomposite 2D array with a Dual Polymer Phase and improved structure.In: IEEE ultrasonics symposium, p 42–45
2. Li W, Sharples SD, Smith RJ, Clark M, Somekh MG (2012) Determination of crystallographic orientation of large grain metals with surface acoustic waves. J Acoust Soc Am 132(2):738–745. doi:10.1121/1.4731226
3. Wilcox PD, Holmes C, Drinkwater BW (2007) Advanced reflector characterization with ultrasonic phased arrays in NDE applications. IEEE Trans Ultrason Ferroelectr Freq Control 54(8):1541–1550
4. Zhang J, Drinkwater BW, Wilcox PD (2008) Defect characterization using an ultrasonic array to measure the scattering coefficient matrix. IEEE Trans Ultrason Ferroelectr Freq Control 55(10):2254–2265
5. Velichko A, Wilcox PD (2010) A generalized approach for efficient finite element modeling of elastodynamic scattering in two and three dimensions. J Acoust Soc Am 128(3):1004–1014
6. Rajagopal R, Skelton E, Lowe M, Craster R (2010) Finite element based hybrid models for ultrasonic NDE using commercial packages. Rev Prog Quant Nondestr Eval 29:662–669
7. Simonetti F (2009) Modern ultrasound imaging: an exciting window into the sub-surface world. Rev Prog Quant Nondestr Eval 28:33–46

Appendix A
The Wave Energy Approach for Computing Group Velocity

In this Appendix an energy approach is used to derive the equation for the group velocity. This derivation is different to that given in Sect. 2.2.3.2 in the main body of this thesis, but yields the same equation. Here the derivation follows Fedorov [1]. Initially an equation that describes the velocity of energy propagation in an elastic plane wave is derived. Then it is shown that this is equal to the group velocity for a lossless medium.

The definition of the velocity vector of the energy flux, V^e, is the time average of the elastic energy flux, S, divided by the time average of the density of elastic energy, E. This is given mathematically as:

$$V_i^e = \frac{\overline{S}_i}{\overline{E}} \tag{A.1}$$

where the bar indicates time average. The elastic energy density is a sum of the strain energy density, W, and kinetic energy density, K:

$$E = W + K \tag{A.2}$$

Strain energy density is given by:

$$W = \frac{1}{2}\sigma_{ij}\varepsilon_{ij} \tag{A.3}$$

Using the relationship $\sigma_{ij} = C_{ijkl}\varepsilon_{kl}$ yields:

$$W = \frac{1}{2}C_{ijkl}\varepsilon_{ij}\varepsilon_{kl} \tag{A.4}$$

and then from the definition of strain:

$$W = \frac{1}{2}C_{ijkl}u_{i,j}u_{i,j} \tag{A.5}$$

Next, the plane wave equation $\left(u_i = Ap_i e^{i(k_i x_i - \omega t)}\right)$ is differentiated with respect to distance, and is substituted into the above equation to give:

C. Lane, *The Development of a 2D Ultrasonic Array Inspection for Single Crystal Turbine Blades*, Springer Theses, DOI: 10.1007/978-3-319-02517-9,

$$W = \frac{1}{2} C_{ijkl} k_i k_j u_k u_l \tag{A.6}$$

The kinetic energy density is given by:

$$K = \frac{1}{2} \rho \dot{u}_i \dot{u}_i \tag{A.7}$$

Differentiation of the plane wave equation with respect to time yields:

$$K = \frac{1}{2} \rho \omega^2 u_i u_i \tag{A.8}$$

Therefore, the elastic energy density is given by:

$$E = \frac{1}{2} C_{ijkl} k_l k_k u_j u_i + \frac{1}{2} \rho \omega^2 u_i u_i \tag{A.9}$$

By referring to Eq. (2.28) from Chap. 2, where the phase velocity was derived, written here again:

$$\rho \omega^2 u_i = C_{ijkl} k_j k_l u_k \tag{A.10}$$

It can be seen that if this equation is multiplied by u_i then the kinetic and strain energy density for a plane wave must be equal. Therefore the elastic energy can be written as:

$$E = \rho \omega^2 u_i u_i \tag{A.11}$$

The elastic energy flux is given by the expression:

$$S_i = -\sigma_{ij} \dot{u}_i = C_{ijkl} \varepsilon_{ij} \dot{u}_i = C_{ijkl} u_{i,j} \dot{u}_i \tag{A.12}$$

Again differentiation of the plane wave equation and substitution yields:

$$S_i = C_{ijkl} k_j \omega u_k u_l \tag{A.13}$$

In high-frequency waves, the energies will vary rapidly with time and therefore it is useful to express them as time-averaged variables:

$$\overline{E} = \int_{-\infty}^{\infty} E\, dt = \int_{-\infty}^{\infty} \rho \omega^2 u_i u_i dt = \frac{1}{2} \rho \omega^2 u_i u_i \tag{A.14}$$

$$\overline{S_i} = \int_{-\infty}^{\infty} S_i dt = \int_{-\infty}^{\infty} C_{ijkl} k_j \omega u_k u_l dt = \frac{1}{2} C_{ijkl} k_j \omega u_k u_l \tag{A.15}$$

Therefore the energy velocity equation is:

$$V_i^e = \frac{\overline{S_i}}{\overline{E}} = \frac{C_{ijkl} k_j u_k u_l}{\rho \omega} \tag{A.16}$$

and because $k_i = kn_i$ and $c = \omega/k$:

$$V_i^e = \frac{C_{ijkl} n_j u_k u_l}{\rho c} \tag{A.17}$$

By comparing the above equation with Eq. (2.47), it can be seen that the group velocity is equal to the energy velocity.

Now it will be proved explicitly that the group and energy velocities are equal for a lossless medium. The original proof was given by Lord Rayleigh [2], however the proof given here follows Sect. 2.7 of Rose [3].

Consider a modulated wave that is formed of a low-frequency carrier wave and a high-frequency component given by:

$$u = 2Acos\left(\frac{1}{2}\delta kx - \frac{1}{2}\delta\omega t\right)\cos(x - \omega t) \tag{A.18}$$

The kinetic energy of this modulated wave is:

$$K = \frac{1}{2}\rho\dot{u}^2 \tag{A.19}$$

The equation of the modulated wave can be differentiated with respect to time using differentiation by parts $(\delta(xy)/\delta z = y\delta x/\delta z + x\delta y/(\delta z))$ to give:

$$\dot{u} = \delta\omega A\cos(x - \omega t)\sin\left(\frac{1}{2}\delta kx - \frac{1}{2}\delta\omega t\right) + 2\omega A\cos\left(\frac{1}{2}\delta kx - \frac{1}{2}\delta\omega t\right)\sin(x - \omega t) \tag{A.20}$$

Squaring the above equation and then assuming all terms with $\delta\omega$ tend to zero yields:

$$\dot{u}^2 \approx 4\omega^2 A^2\cos^2\left(\frac{1}{2}\delta kx - \frac{1}{2}\delta\omega t\right)\sin^2(x - \omega t) \tag{A.21}$$

Therefore the kinetic energy is approximated by:

$$K \approx 2\rho\omega^2 A^2 cos^2\left(\frac{1}{2}\delta kx - \frac{1}{2}\delta\omega t\right) sin^2(kx - \omega t) \tag{A.22}$$

The energy is averaged over n number of time periods, T. The low-frequency term changes very slowly so, over a few cycles of the high-frequency wave, is effectively constant. Therefore, the low-frequency term can be removed from the integral. The time-average of high-frequency term is given by:

$$\frac{1}{nT}\int_0^{nT} sin^2(\delta kx - \delta\omega t)dt = 0.5 \tag{A.23}$$

Therefore the time-averaged energy density equation is:

$$\overline{K} \cong \rho\omega^2 A^2 cos^2\left(\frac{1}{2}\delta kx - \frac{1}{2}\delta\omega t\right) \tag{A.24}$$

The velocity at which the energy propagates is therefore given by:

$$V_i^e = \frac{\partial\omega}{\partial k} \tag{A.25}$$

This equation is the same as the group velocity Eq. (2.7) and therefore proves that the group velocity is the velocity of energy propagation in a lossless matxerial.

Appendix B
Analysis of Finite Element Accuracy for Beam Amplitude Modelling

In this Appendix, the accuracy of finite element (FE) models, for the computation of beam amplitudes from a point-force on an elastic half-space, is investigated. This is done by using FE models with different mesh densities and comparing the results to the analytical result for an isotropic medium.

The analytical solution for the far-field beam amplitude, D, of the longitudinal wave mode from a normal point-force acting on surface of an isotropic half-space at an angle, θ, is given by Miller and Pursey [4] as:

$$D(\theta) = \frac{\left((c_l/c_s)^2 - 2\sin^2\theta\right)\cos\theta}{F(\sin\theta)} \tag{B.1}$$

where:

$$F(\zeta) = \left(2\zeta^2 - \left(\frac{c_l}{c_s}\right)^2\right)^2 - 4\zeta^2(\zeta^2 - 1)^{1/2}\left(\zeta^2 - \left(\frac{c_l}{c_s}\right)^2\right)^{1/2} \tag{B.2}$$

The longitudinal and shear velocities, c_l and c_s, can be found from the Young's modulus, E, and the Poisson's ratio, v, for an isotropic elastic solid using:

$$c_l = \left(\frac{E(1 - v)}{\rho(1 + v)(1 - 2v)}\right)^{\frac{1}{2}} \tag{B.3}$$

$$c_s = \left(\frac{E}{2\rho(1 + v)}\right)^{\frac{1}{2}} \tag{B.4}$$

Figure B.1 shows a schematic diagram describing the FE model. The model was produced in ABAQUS 6.6 using eight-noded cuboid isotropic elements. To reduce the model size, symmetric boundary conditions were used about the x-z and y-z planes at the origin. This was achieved by forcing the displacements in the x direction to zero for nodes that were located at x = 0. Likewise, y displacements for nodes at y = 0 were forced to zero. A single cycle force with a centre frequency of 10 MHz was applied at the origin. The longitudinal beam amplitudes were obtained from the collected time-history traces of monitoring nodes, which

C. Lane, *The Development of a 2D Ultrasonic Array Inspection for Single Crystal Turbine Blades*, Springer Theses, DOI: 10.1007/978-3-319-02517-9,

Figure B.1 Schematic diagram showing FE model

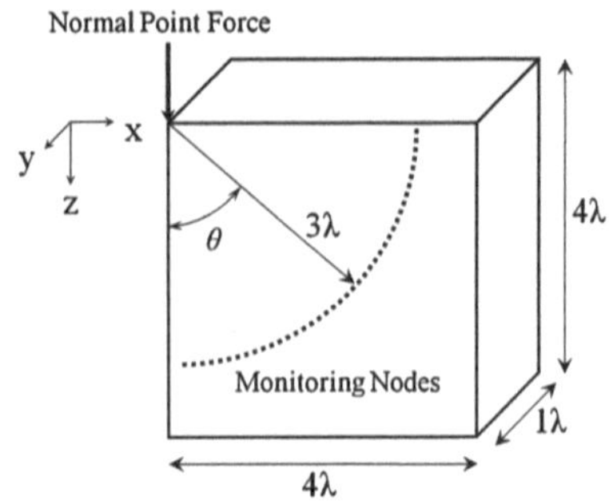

Table B.1 Properties of the isotropic solid material used for the FE analysis

Material property	Value
E	221 GPa
ν	0.3
ρ	8120 kg/m^3
c_l	6.053 km/s
c_s	3.235 km/s

were located nominally at three longitudinal wavelengths from the point-force at angles from 0 to 90°, in 1° increments. The FE model was used to compute the beam amplitudes in this way using mesh densities of 30, 40, 50, 60 and 100 FE elements per longitudinal wavelength. Table B.1 shows the material properties of the isotropic solid used in this analysis.

Figure B.2 shows the computed far-field longitudinal beam amplitude using the FE model with different mesh densities and the analytical model. It can be seen that at high angles there is a large discrepancy between the analytical and FE models. This is due to the presence of surface waves (in this example surface skimming longitudinal waves) that occur in the FE model but are not included in

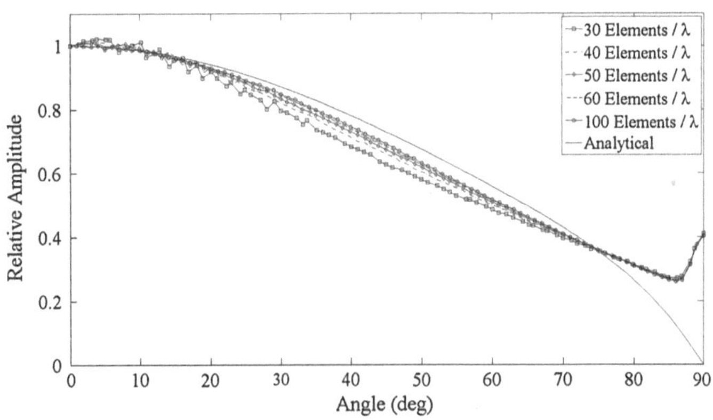

Fig. B.2 Graph showing the relative beam amplitudes found using the FE models with different mesh densities and the analytical model

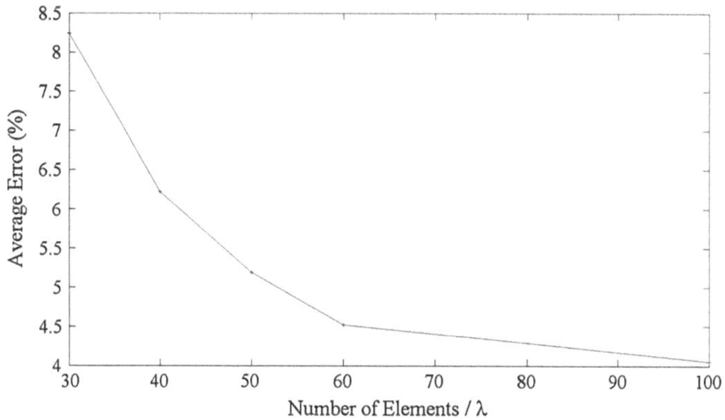

Fig. B.3 Graph showing the average error between the FE model at different mesh densities and the analytical solution. Beam amplitudes that were affected by the surface wave interference were excluded from the error calculation

the analytical analysis. In regions away from the surface wave interference, the difference between the results from the FE models and the analytical model decreases, as the number of elements used in the FE models increases. This is also seen in Fig. B.3 which shows a plot of average error between the FE and analytical models as the FE mesh density increases. The FE results can be seen to converge with increasing mesh density.

The results, however, do not converge on zero error. It was thought that this was due to the fact that the analytical equations are asymptotic solutions and are therefore only exact at the very large distances away from the source. The FE results were computed using a relatively small distance of three wavelengths. Therefore, 2D FE models were created, using a mesh density of 100 elements per

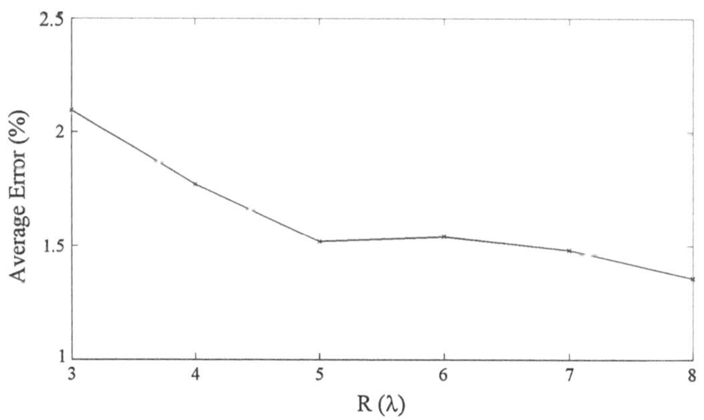

Fig. B.4 Graph showing the average error between the 2D FE model using different distances between the source and monitoring nodes (R) and the analytical solution. A mesh density of 100 elements per wavelengths was used

wavelength, to examine the effect of distance, R, between the source and monitoring nodes on the measured beam amplitude. As Fig. B.4 shows, the average error between the results from the FE and analytical model reduces as the distance of the monitoring nodes from the point-force increases.

Appendix C
The Gaussian Curvature of a Surface

In this Appendix the Gaussian curvature of a surface is defined. The Gaussian curvature of the slowness surface is used in the asymptotic anisotropic Green's function in the main body of the thesis to compute the beam amplitude from a point-force acting on the surface of an anisotropic half-space.

The curvature of a surface is essentially a measure of how much the surface deviates from being flat [5]. One way of measuring the curvature of an arbitrary surface in a quantitative way is by computing the Gaussian curvature. The Gaussian curvature, K, of a surface is the product of the principle curvatures, κ_1 and κ_2:

$$K = \kappa_1 \kappa_2 \tag{C.1}$$

The principle curvatures are the maximum and minimum curvatures at a point on the surface. To illustrate, consider the curve shown in Fig. C.1. At the point P on the surface of the curve the directions of maximum and minimum curvature are labelled κ_1 and κ_2 respectively. The direction of maximum curvature can be seen to be the direction in which the curve of the surface increases most rapidly. Conversely, the direction of minimum curvature is the direction in which the curve decreases most rapidly.

For a curved surface that is given in the form:

$$r = r(u, v) \tag{C.2}$$

where r is a position vector describing a point on the surface, whose components can be given by Cartesian coordinates (x, y, z), uniquely defined by curvilinear coordinates u and v, it can be shown [6] that the Gaussian curvature is given by:

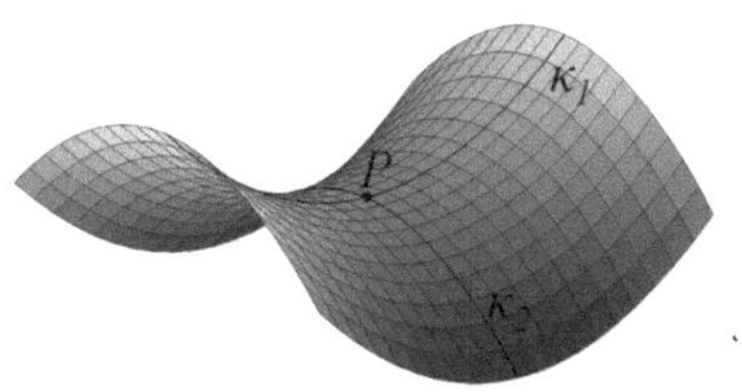

Fig. C.1 Curve with principle curvatures indicated in the *bold lines* and labelled κ_1 and κ_2 for point P

C. Lane, *The Development of a 2D Ultrasonic Array Inspection for Single Crystal Turbine Blades*, Springer Theses, DOI: 10.1007/978-3-319-02517-9,

$$K = \frac{LN - M^2}{(EG - F^2)} \tag{C.3}$$

where:

$$L = \frac{\partial^2 r}{\partial u^2} \cdot n \tag{C.4}$$

$$N = \frac{\partial^2 r}{\partial u \partial v} \cdot n \tag{C.5}$$

$$M = \frac{\partial^2 r}{\partial v^2} \cdot n \tag{C.6}$$

$$E = \left(\frac{\partial r}{\partial u}\right)^2 \tag{C.7}$$

$$G = \left(\frac{\partial r}{\partial v}\right)^2 \tag{C.8}$$

$$F = \frac{\partial r}{\partial u} \cdot \frac{\partial r}{\partial v} \tag{C.9}$$

and n is the unit vector normal to the curve given by:

$$n = \frac{\frac{\partial r}{\partial u} \times \frac{\partial r}{\partial v}}{\sqrt{EG - F^2}} \tag{C10}$$

It is worth noting that there are a number of pre-programmed MATLAB functions available for free download on the MATLAB Central File Exchange [7] that can compute the Gaussian curvature of an arbitrary surface.

Appendix D
The Development of a Single Element Probe for the Inspection of Turbine Blades

The development of a single element probe for the inspection of trombone cracking in the root section of the HPT1 blade is described in this appendix. The single element probe was developed so a direct comparison between this conventional ultrasonic probe and the ultrasonic array probe described in Chap. 5 of this thesis could be made.

The single element probe was manufactured by placing a 10 MHz ultrasonic transducer on the platform of a turbine blade, which contained a 3mm long through-going slot in the root section, with a minimal layer of couplant between the transducer and the blade platform. This is shown schematically in Fig. D.1. The element was operated using an EPOCH 4 flaw detector manufactured by Olympus Inspection and Measurement Systems (Tokyo, Japan). The position and orientation of the ultrasonic transducer was varied using a micro-positioner until the signal from the slot was maximised. The gain of the flaw detector was increased until the signal from the slot was 100 % of the screen-height. The transducer position was then fixed by surrounding the transducer with epoxy-based glue. This glue, once set, formed the housing of the ultrasonic probe and conformed to the shape of the turbine blade so that the probe could be positioned in the same location on other blades.

Figure D.2 shows examples of A-Scans from the ultrasonic flaw detector using the single element probe to inspect a turbine blade with a 3 mm long through-going slot and a blade without a slot. The slot signal can be seen clearly at 3.5

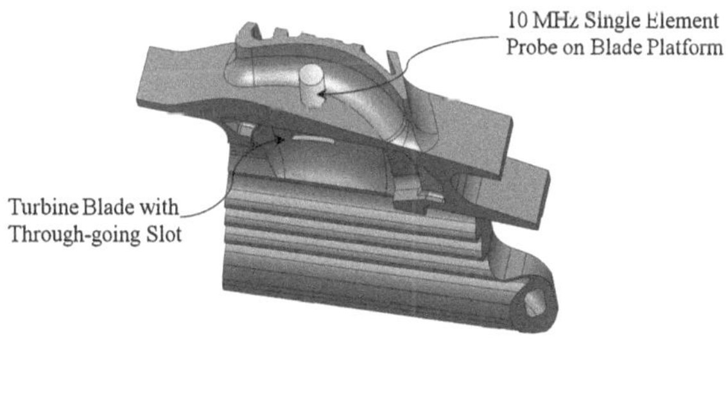

Fig. D.1 Schematic diagram showing location of the 10 MHz single element probe on the blade platform. Image courtesy of Rolls-Royce plc

C. Lane, *The Development of a 2D Ultrasonic Array Inspection for Single Crystal Turbine Blades*, Springer Theses, DOI: 10.1007/978-3-319-02517-9,

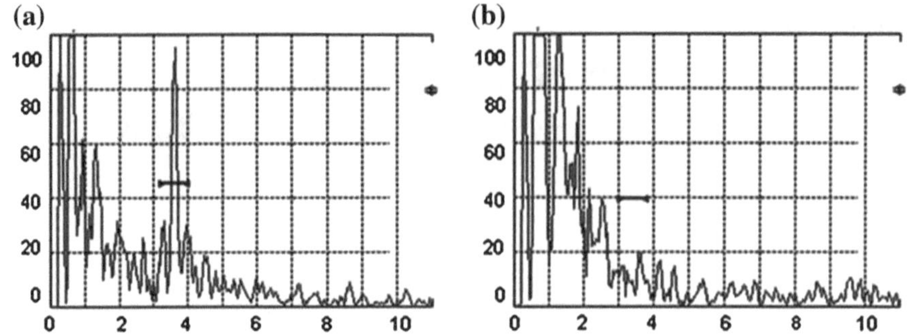

Fig. D.2 Example images of A-Scan from the single element inspection of (**a**) a turbine blade with a through-going 3 mm long slot and (**b**) a blade with no slot. Each division on the x-axis represents a time-of-flight of 0.75 μs

divisions on the x-axis of Fig. D.2a. It can be seen from Fig. D.2b that the signal from the slot-free blade at the same location was approximately 20 % screen-height. This signal was mainly caused by reflections off features of the turbine blade geometry and to a lesser extent by electrical noise.

Appendix E
Probability of Detection

Probability of detection (PoD) analysis was applied to the inspection of the HPT1 blades with the ultrasonic phased array probe in Chap. 6 of the main thesis. From this analysis a PoD curve was produced for cracks with different through-going extent. In this appendix, the use and derivation of PoD curves is explained in more detail.

PoD analysis is a quantitative method used to measure the sensitivity of a NDE inspection [8, 9, 10]. In essence it determines the smallest defect size that one particular inspection will reliably detect. There are two types of data that can be used in PoD analysis: *Hit or Miss* data or *Signal Response* data. Hit or Miss data is used for NDE techniques that either detect the defect or not, for example with fluorescent penetrant inspection. Whereas Signal Response analysis is used for inspection techniques that give a measurable response from a defect such as eddy current inspection or ultrasonic testing. This project was concerned with ultrasonic testing therefore Signal Response analysis is most appropriate and will be considered in more detail.

The overall scheme of PoD analysis is shown schematically in Fig. E.1. Conventional nomenclature for Signal Response analysis defines the signal response as $\hat{a}$ and the defect size as a; hence this analysis is sometimes referred to as *a-hat* versus a analysis. Initially a relationship is found between the defect size

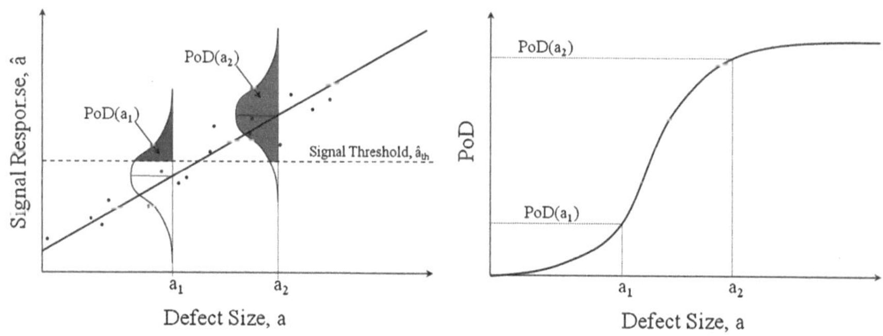

Fig. E.1 Schematic diagram showing the PoD concept

C. Lane, *The Development of a 2D Ultrasonic Array Inspection for Single Crystal Turbine Blades*, Springer Theses, DOI: 10.1007/978-3-319-02517-9,

and signal response so that the expected response at any given defect size can be predicted. The predicted response has an error associated with it as the experimental response will vary from the predicted response due to other inspection variables. This variation can also be modelled. A signal threshold is then set to be greater than the maximum noise level. At every defect size the probability that the signal response is greater than the threshold is computed and plotted on a PoD curve.

The probability of detecting a defect of a given size, a, is the probability that the signal response from the defect, $\hat{a}$, is greater than a pre-determined threshold value, $\hat{a}_{th}$. This can be stated mathematically as:

$$PoD(a) = P\{\hat{a} > \hat{a}_{th}\} \quad \text{(E.1)}$$

The value of the threshold is determined by the noise of the inspection system. To minimise the false-call rate, the threshold value must be greater than the noise. It is typical to set the threshold to be twice the maximum value of measured noise from a defect-free test specimen.

To produce a continuous PoD curve (i.e. a PoD value for a continuous range of defect sizes) the signal response must be modelled using parameters computed from discrete measurements. For example, many empirical studies have shown [11] that there is a linear relationship between the logarithm of a and the logarithm of $\hat{a}$ hence:

$$y = \beta_1 + \beta_2 x + \delta \quad \text{(E.2)}$$

where:

$$y = \log(\hat{a}) \quad \text{(E.3)}$$

and:

$$x = \log(a) \quad \text{(E.4)}$$

The variables β_1 and β_2 are experimental constants and δ is the random variability of the data. β_1 is the measured response for $x = 0$ and β_2 is the constant of proportionally for the relationship. These parameters can be computed from the experimental data using curve fitting analysis. The log-log relationship may not necessarily be the best fit for all inspections. Therefore, a range of relationships should be tested on each data set to find the most appropriate.

The random variability of the data, δ, reflects the fact that defects of the same size can give a range of signal responses due to other changes in the defect variables (e.g. crack roughness, orientation etc). It is assumed that δ is normally distributed with zero mean and constant standard derivation, σ. The normal distribution function, ϕ, can be expressed as:

$$\phi(\xi) = \frac{1}{\sqrt{2\pi\sigma^2}} e^{-\frac{(\xi - \mu)^2}{2\sigma^2}} \quad \text{(E.5)}$$

where μ is the mean and represents the position of the peak value. The term $1/\sqrt{2\pi\sigma^2}$ ensures that the area under the distribution function is equal to unity. The value σ^2 is named the variance and can be computed from the residuals of the experimental response data and the expected signal response.

Referring to Figure E.1, it can be seen that the probability of detecting a defect of size a, is computed by integrating the normal distribution function between the detection threshold and infinity. The expected signal response from a defect of size a is given by $\beta_1 + \beta_2 x$. As a result, the peak of the normal distribution function (i.e. its mean) is equal to this value. Therefore, this can be expressed as:

$$PoD(a) = \frac{1}{\sqrt{2\pi\sigma^2}} \int_{\log(\hat{a}_{th})}^{\infty} e^{-\frac{(\xi - (\beta_1 + \beta_2 x))^2}{2\sigma^2}} d\xi \tag{E.6}$$

This integration cannot be performed analytically, however, the solution can be expressed using the well-known cumulative distribution function, Φ. The cumulative distribution function is defined as

$$\Phi\left(\frac{z-\mu}{\sigma}\right) = \frac{1}{\sqrt{2\pi\sigma^2}} \int_{-\infty}^{z} e^{-\frac{(\xi-\mu)^2}{2\sigma^2}} d\xi \tag{E.7}$$

As the normal distribution function is symmetric and its area is unity, it is simple to show that:

$$\Phi(-z) = 1 - \Phi(z) \tag{E.8}$$

Therefore:

$$\frac{1}{\sqrt{2\pi\sigma^2}} \int_{z}^{\infty} e^{-\frac{(\xi-\mu)^2}{2\sigma^2}} d\xi = 1 - \Phi\left(\frac{z-\mu}{\sigma}\right) = \Phi\left(-\frac{z-\mu}{\sigma}\right) \tag{E.9}$$

Returning to Eq. (E.6) it follows that:

$$\frac{1}{\sqrt{2\pi\sigma^2}} \int_{\log(\hat{a}_{th})}^{\infty} e^{-\frac{(y - (\beta_1 + \beta_2 x))^2}{2\sigma^2}} dy = \Phi\left(-\frac{\log(\hat{a}_{th}) - (\beta_1 + \beta_2 x)}{\sigma}\right) \tag{E.10}$$

Therefore, the probability of detecting a defect of a given size is:

$$PoD(a) = \Phi\left(\frac{x - \hat{\mu}}{\hat{\sigma}}\right) \tag{E.11}$$

where:

$$\hat{\mu} = \frac{\log(\hat{a}_{th}) - \beta_1}{\beta_2} \tag{E.12}$$

and:

$$\hat{\sigma} = \frac{\sigma}{\beta_2} \tag{E.13}$$

One approximate way to model the cumulative distribution function and hence produce a PoD curve is to use the inverse-logit function:

$$\Phi(\xi) \approx \frac{1}{1 + e^{-\xi}} \tag{E.14}$$

Alternatively, the cumulative distribution function can be computed to a greater accuracy using the Taylor expansion:

$$\Phi(\xi) = \frac{1}{2}\left[1 + \frac{2}{\sqrt{\pi}}\sum_{n=0}^{\infty}\frac{(-1)^n(\xi/\sqrt{2})^{2n+1}}{n!(2n+1)}\right] \tag{E.15}$$

Therefore, by using the statistical analysis presented above on experimental data, the probability of detecting a defect of any size can be computed and displayed as a PoD curve.

It is common to quote the smallest defect size that can be found with a PoD of 90 %. This is usually denoted as $a_{90.}$ In addition, the PoD result and a_{90} value is often expressed with an associated confidence. This confidence reflects the fact that a limited sample size of defects is used to conduct the PoD analysis. Therefore, if a different sample of defects is used, the PoD results may change. It is typical to express the a_{90} value with a 95 % confidence level. This is denoted $a_{90/95}$. Essentially what this means is that if an infinite number of defects are used, a true value of a_{90} will be obtained. However, if the PoD analysis is performed 100 times with a different but a finite sample of defects, the computed $a_{90/95}$ value will be greater than the true a_{90} value 95 times out of the 100.

The computation of the $a_{90/95}$ value requires further statistical analysis [9]. However, the analysis presented above will produce an a_{90} value that is approximately half the time greater than the true a_{90} value and half the time less. Therefore, this PoD result would have an associated confidence of 50 %.

References

1. Fedorov FI (1968) Theory of elastic waves in crystals. Plenum Press, New York
2. Rayleigh L (1926) Theory of Sound, vol 1. Macmillan and Co., Ltd., London
3. Rose JL (1999) Ultrasonic Waves in Solid Media. Cambridge University Press, Cambridge
4. Miller GF, Pursey H (1954) The Field and radiation impedance of mechanical radiators on the free surface of a semi-infinite isotropic solid. Proc Roy Soc A 223(1155):521–541
5. Stillwell J (1992) Geometry of Surfaces. Springer, New York Universitext
6. Lord EA, Wilson CB (1986) The mathematical description of shape and form. In: Bell (GM) Mathematics and its Applications. Wiley, New York
7. Mathworks (2011) MATLAB Central File Exchange. [cited 2011 06 May]; Available from http://www.mathworks.com/matlabcentral/fileexchange/

8. Annis, C. *Statistical Engineering*. 2011 [cited 2011 23 April]; Available from: http://www.statisticalengineering.com
9. Nondestructive Evaluation System Reliability Assessment. Vol. MIL-HDBK-1823. 1999: USA Department of Defence Handbook
10. Georgiou GA (2006) Probability of Detection (PoD) Curves: Derivation, applications and limitations. Vol. research report 454. Health and Safety Executive Books, Suffolk
11. Berens AP (1989) NDE reliability data analysis, in Metals handbook. ASM International, pp 689–701

FSC
www.fsc.org
MIX
Papier | Fördert
gute Waldnutzung
FSC® C083411

Kolibri 360 GmbH
Ferdinand-Jühlke-Straße 7
99095 Erfurt, Deutschland
produktsicherheit@kolibri360.de